MY BROKEN SOLDIER

THE UNTOLD STORY OF LIFE BEYOND THE FRONT LINE

KAREN PAGE

My Broken Soldier

978-0-6451016-0-7 PRINT
978-0-6451016-1-4 EPUB
978-0-6451016-2-1 MOBI

NATIONAL LIBRARY OF AUSTRALIA
A catalogue record for this book is available from the National Library of Australia

To contact the author, or to purchase additional copies of this book

www.kazpage.com.au

Proofreading by Alan Tasker
Editing by Phaedra Pym : awaywithwords.net.au
Produced and printed for the author by exlibris.com.au

Acknowledgements

I once asked Matt in a fit of despair, 'How do I move forward and how do I bring awareness to this?' His response was simply, 'Write a book.' He planted the seed and he supported and encouraged me all the way. Matt, you are, you have always been, and you will forever be, my hero. Thank you for believing in me.

This story could have had a very different outcome were it not for a key group of friends who supported us on different levels, and you are all in the book. To each and every one of you, I am so grateful.

This book would not have seen the light of day without the help of my editor, Phaedra Pym. This was a first for me and it started as a hot mess. But with Phaedra's insight, experience and guidance we have turned it into something that I am so very proud of, a book that I hope will help some, if not many, people. Thanks for pushing me further than I thought possible and for keeping me focused to the end.

To publish this book, to help me get it to you, we needed to find a publisher that saw me, understood how new this was to me and guided me through with a 'whatever it takes' attitude. I could not have found a better partner to publish this book for us. The team at Exlibris are a perfect match and I can't thank Eric and his team highly enough… You really have made this part of the journey easy.

Contents

Introduction

Matt once told me, 'You and I were never childhood sweethearts, but we will spend the rest of our lives making up for it'. He was nineteen, not long out of school, a time when the whole world is laid out in front of you. But strangely he gave it all up for his best friend, me. He was ten years my junior, we had disapproving families and he hadn't really worked out where he fitted in the world yet. To say it wasn't all smooth sailing would be an understatement. By the end of it we would barely have the clothes on our backs. Somehow, despite making no sense in any conventional means and at times, no matter how hard we fought it, the Universe kept rerouting our paths back to one another.

At the beginning, this might sound like your typical romantic piece of pulp fiction. But it is so much more than that. Our story transcends financial ruin, family politics, the loss of babies and people we love. We survived break-ups, all-consuming careers, two wars, life-changing injuries, financial ruin (twice), utter betrayal by those who were supposed to have our backs, the shortcomings and injustice of a system too big to fight, and our high-paced maniacal lifestyle that only an ADHD-riddled stockbroker could match for work ethic and ambition.

So who am I? And why have I written this book? I am Karen Page – Kaz to those closest to me – a wife, a mum, a worker, a daughter and a friend. After a fairly turbulent upbringing as the second child in a blue-collar Aussie family from suburban Sydney, two failed marriages and with two young daughters in tow, I finally found a man who became my best friend, my rock and my lifelong partner in crime. But he also became a soldier. And in that moment, I became a soldier's wife.

Today I am still married to my soldier. I am still in love with him; he is still my rock, my best friend and he still makes me laugh like no one else can. But my husband is not the same man I married all those years ago. Today he is in pain from wounds and damage no one else can see. My husband is a broken soldier. Being a broken soldier is not a badge of honour, however it should be. He put on a uniform, went to war twice and put his life on the line, sacrificing himself and his family to protect a country that wasn't even his own.

After all we have endured, after all we survived, I decided to put pen to paper and write our story. I've written this story for me, for him and for our family. But I've also written it for the unseen, the unacknowledged, the quiet sufferers of PTSD, anxiety, adjustment disorder, depression, substance abuse, blast trauma, bullet wounds, and all the other wounds no one sees. Our story is also for all those who, like me, love the broken, the damaged, including families of soldiers who are either directly affected by their time spent at war or who never returned, in a literal sense or a psychological one.

This book is written on the backs of unpayable bills, endless doctor's reports, death notices, and thousands of tissues that have caught countless tears, never to be spoken of in the light of day. It is important to know that there are so many more casualties to war than those created in a war zone, casualties that will never be treated, never observed, never given a voice. This isn't a story about lifesaving medicine and we're not celebrities. This is our life – and mine, as the wife of a soldier who returned a different man, and as the wife who supports the sufferer and keeps the family going. We are one of the lucky couples. To be honest, we probably wouldn't have been so lucky were it not for some amazing people who helped, supported and offered respite at crucial moments. Many aren't so lucky.

My hope is that after reading this book, you will come to understand and have a little more respect and appreciation for what the men and women behind the men and women in uniform – any uniform – endure, especially when one or more of the unspoken mental health-related illnesses becomes deeply entrenched in their lives, like a wound that never quite heals.

There is much that needs to change in both the Australian Defence Force and the Department of Veterans' Affairs to stop the steady and constant flow of suicide within our returned troops and to better help the families supporting them. I am but one voice. Hopefully through this book, my voice will resonate loudly enough to help change come faster, smarter and with far greater empathy than exists today. Unfortunately change is hard, and a massive ship like the Australian Defence Force turns slowly. This is why I think it is critically important to share this story, a story that reveals what happens when a soldier goes to war and leaves a wife and children behind; a story that explains, warts and all, what happens when a soldier is seriously injured and the agonising journey their partner goes through to pull everyone through.

I truly believe it is so important for people to understand that when a soldier goes to war, the family they leave behind experience a type of war of their own. War affects us all. So, next time you come across a broken soldier – or anyone in uniform – you might have a better insight, not only into what they endure, but also the journey taken by their partner, kids, friends and other family back home, as they walk a treacherous path of their own. Acknowledge them, respect them, and take a moment to wonder or ask how they are going.

Some say this is Matt's story through my eyes, but it isn't; this is our story, the untold story of life beyond the front line for those who serve and the loved ones who support them. Matt's personal reflection and perspective on key aspects I have shared are interspersed throughout the book. While our story involves tragedy, despair and heartache, it is ultimately a story of resilience, sacrifice, love, inspiration and hope for the future.

Prologue

Matt often says that my father spoilt me: that I was probably a petulant child and this made me a petulant adult. That may be true, to a degree, but I think we both agree that my father had no clue about emotion or love, and this was the only way he knew how to show he cared. My father was also a hard man. While he gave us 'things', he didn't give emotional support. He didn't know how to convey his affection towards my brother and me; he knew how to show us anger, disappointment and even disdain, but never pride or encouragement or love. No matter what I achieved through primary or high school, it was never enough. I grew up feeling like I was a constant disappointment, never quite good enough. So I rebelled, all through my teens and beyond. I tried every drug possible at the time and got my first of many tattoos before I was legally allowed to.

You're drawing a mental picture of me right now, aren't you? And you're probably not far off the mark. Just remember though, the cover doesn't always fully describe the book.

I hated school, and all too often school hated me back. It made for some really tough years during which I had no real friends. I bounced from group to group, never really knowing where I belonged or how to fit in. Later in life some of my old school 'friends' became some of my closest adult friends. It's funny how the world turns you full circle without you even realising it.

As I navigated my way through the teenage years, I made many mistakes and put myself in some unthinkable situations. Some led to disastrous outcomes, like being raped at fourteen and not having a single soul to turn to. I felt at the time that I had to hide what happened to me, even from my parents. I had no idea about the long-term emotional ramifications of what happened that night; I learned that lesson much later in life. I just knew at the time that the implications of telling my hot-under-the-collar father were far more frightening than keeping it to myself. And I didn't want to burden my mother with it.

I spent my life watching my father cheat on my mother. And I watched Mum bury this hurt in the bottle and a solitary life at home, doing only the things that gave her enjoyment away from home, like playing tennis, going fishing and helping everyone except her own children. The truth is, my parents didn't know me, nor did they seem to want to really know me. My brother and I were to be seen and not heard, fed and clothed but not shown any love beyond material possessions. There was no real love in our home. So I grew up hard and fast, and in later years, I became a stubborn, pain-in-the-arse adult who would do shit whether my parents liked it or not – with or without them. But I learned to endure. My father eventually discovered that doing things like building horse stables and sheds and spending time with me and my horses was really the only way he could be a part of something with me in my adult years. Eventually it worked for both of us.

Today, I am still stubborn, I can be arrogant, and yes, I absolutely have a petulant streak. I don't take bullshit from anyone. But I am loyal to the core, and I am honest. I will tell you the truth no matter how that looks. I am strong, I am resilient, and I am generous. I will give you the shirt off my back if you need it – on one condition – that when you are back on your feet and able, you give your shirt to the next person in need. I truly believe my strength, stubborn streak and even my petulance have helped me survive and get through some of the toughest times of my life.

CHAPTER 1
Age is just a number

I was twenty-eight, married for the second time and living on the Central Coast with my husband and two daughters, Jessica (Jess) and Ayleigh. My daughters both loved my second husband. My eldest, Jess, came from my first marriage and didn't really have a lot to do with her biological father. Unfortunately by this point, my second marriage was also falling apart, making me conclude that I sucked at marriage. The thought of having to go through yet another shitty, vicious divorce, this time with two kids to protect and navigate through it all, was both heartbreaking and terrifying. To everyone else, however, we were perfectly happy and everything was fine. We had plenty of friends and we socialised a lot on weekends, I think largely to prevent us having to spend too much time together. One particular weekend, we were invited to a friend's house for dinner. Our friend explained that there were some other people coming along that he wanted us to meet. One of those people was Matt.

Matt was just eighteen at the time with a great sense of humour, a certain cockiness about him and a gorgeous smile that seemed to light up the room. Everyone got along so well that night that I knew we would all see each other again. So a few weeks later I decided I would return the favour and invite everyone over to our place for dinner, including Matt, who everyone called 'Pagey', and the girl he had brought with him to the last dinner party. I thought Pagey was a funny name for a boy (I didn't know his first name was Matt at the time), but just went with it. His girlfriend was nice enough and they seemed happy. Despite the mix of ages in the group, we all got along so well that we started hanging out more regularly. My best friend,

Justine (Jussy), was the only person in the group who knew the truth about my marriage and that it was on its last legs.

During that dinner at my house, I couldn't help myself and asked Pagey what his last name was (I will admit to being a bit of a sticky beak). He responded with 'Page'. 'Pagey Page! What the…?' Unfortunately, not only did my brain go there, my mouth promptly followed! I'm pretty sure he thought I was dumber than a shit carter's horse in that moment. But he just laughed and clarified that his name was Matthew Page. From that moment on he was and still is Matt to me. I never did get the hang of calling him Pagey.

Somehow, Matt and I just gelled. We became great mates and started spending a lot of time together. We could talk for hours and hours, about anything, everything and nothing. The more time we spent together, the more we discovered about each other, and even though Matt was ten years younger than I was, it didn't feel like there was any difference at all. He was a lot more mature than any other eighteen-year olds I knew and maybe I acted a lot younger than a twenty-eight year old should, who knows? Anyway, we just seemed to have a lot in common and enjoyed hanging out together. I didn't realise at the time how much his friendship would help me get through the next difficult phase in my life.

Matt had his own issues and he seemed comfortable opening up to me. I listened to him talk about his family, including his grandparents, how much he missed his Pa and the impact his passing has had on him. On the surface, Matt was raised in a pretty normal white-collar family of four on the New South Wales Central Coast – his mum, dad, younger brother and himself. But scratch the surface and you start to realise his family was anything but normal, although, given my own upbringing, I'm not sure I am suitably qualified to define a 'normal' family. Matt's parents' recollection of his upbringing is very different to his, but I have learned over the years that when it comes to humans and our ability to recall a period in time, an experience or a memory, two people who have lived through the exact same day together can have completely different memories of that day. For instance, in writing this book, I discovered that my recollection of growing up differs from my brother's, and my mother's account is also different. The same was clearly true of Matt's family. This doesn't make our own personal journey any less real, our memories any less significant; it just means that, as humans, we all take something different away from each experience.

Matt's memory of his childhood is not filled with joy and happiness; it is filled with bitter memories and pain, both physical and mental. I'm sure there were good times as well, but the hard stuff is what shapes us; it's what gives us the foundation to help us become adults. Pain is what we remember first.

MATT'S PERSPECTIVE...

To say that my childhood was typical would be a lie, but to say it was horrific would also be a lie. One thing I can say is that if you were to speak to a child psychologist from any school of thinking, they would not recommend the myriad issues that swirled in the miasma of which I set my roots and began to do my best to grow.

One of my two earliest memories was of my mother and my Nan having a heated argument outside a church over (to the best of my recollection) me. I remember it being sunny, so sunny it hurt my eyes. I recall it was a very heated argument and, upon reflection, it was more like a power struggle about ownership of me than anything else. That power struggle went on for years. The other early memory I have is of the 'treatments to my boy parts' I supposedly had to undergo, according to a 'doctor' who assured my mother that everything was perfectly normal in what he was doing to me and that I would need regular 'examinations' for the 'treatment', which was all part of an early intervention. To be honest, I was young and didn't really give it much thought at the time. It wasn't until much later in life that I questioned the cause and effect. Looking back now it was dodgy as fuck, but I don't think about it much, nor do I ever really discuss it. Neither of these memories, the kinds of memories that have the potential to damage a person, or others, later in life that I experienced at the hands of adults in my world that should have done better, defined me. They could have. But I chose not to use them as reasons for why things might not be turning out my way. I don't look to my past for excuses about today.

Neither my Nan nor my mother had great lives. Nan was raised by her grandmother and times were tough back then. So as soon as was possible, Nan was sent out to work and was treated like a millstone around her grandmother's neck until she married my Pa. Unfortunately they couldn't conceive, making Nan the millstone of yet another family. When the opportunity to adopt a baby girl came along, they jumped on it. However they soon discovered that this bundle of joy came with Foetal Alcohol Syndrome and was in detox. This was yet another misdemeanour in my Nan's life, so she did what many others in her situation would do: she dumped all the hurt and blame and control, as well as need for attention and fulfilment squarely onto this already burdened baby, who would grow up to be my mother. My mother endured much and was given little of anything that a loved child should be given. Not only suffering a learning disability and several incorrectly treated autoimmune issues, she also suffered the torment of what I would come to believe was Munchausen Syndrome by Proxy at the hands of her mother, my Nan.

Despite all that she endured, my mother met my father, who came from what I would call a reasonably normal, happy suburban family, and somehow they managed to stumble their way into marriage and stay married for the long haul. Needless to say, this doesn't make for the most nurturing environment and these women, who never learned how to nurture, how to show empathy or love, were never able to teach it either. By the time my mum had children, she was way out of her depth, barely able to cope with her own psychological trauma let alone worry about what her actions or behaviours might do to her own children. Dad travelled a lot for work, although deep down I believe this was as much for his own sanity as it was for the money. To be honest, I can't say I really blame him; it was no picnic at home. Dad always held what seemed like good, strong professional jobs that should have paid well, however we never had any money and often couldn't afford the things we needed when we needed them. My brother and I regularly had our savings accounts garnished by our mum, not that this was ever admitted to.

Today, I sometimes think my mum lives in a reality that is just slightly off-centre to ours, not completely parallel or perpendicular, just slightly off. She also has a bunch of little 'ticks' that she will carry with her till the day she dies, and I have learnt that my job now is to simply limit the effect they have on my children while just letting her be a Grandma to them.

In addition to those dubious 'treatments', I spent the first two years of my life unable to be bathed in water. Later on this would evolve into sensitivity to a host of normal everyday products as well as chronic asthma. I was on a cocktail of corticosteroids and various other drugs that came with a range of unpleasant side effects. They made me sweat, they made me thirsty, they gave me body odour, had me itching and wheezing at times, numb and dizzy at others. There was no joy in any of this.

Once I got to high school I was awkward and not particularly popular. In hindsight, given my health and medication side effects, it's not hard to understand why. As if being a social outcast wasn't enough, I was then hit with a growth defect missed in my early childhood, which required me to have my femur pinned to prevent the joint from becoming necrotic. This meant I had to spend four months in severe physical pain, basically limited to minimal movement, which resulted in significant weight gain. Then the surgery had to be repeated twelve months later! By this stage I had learned to just go with it.

The next few years were the usual whirlwind of teenage years, however I did meet a man who would become a kind of mentor to me. He gave me a place to hide, a place to learn and a place where I was liked for just being me. Ronnie was a joiner and, at the time, he was also an addict (you name the drug and he was that type of addict). I found out through a friend that he was looking for a workshop assistant. It was the perfect afterschool job, five dollars an hour and all the cigarettes I could smoke if I wanted them. Ronnie was what you could call a 'renaissance man'. He could build anything, he seemed to know everything there was to know about the world, he was an avid sailor, a mad scientist, an artist and, most significantly of all in my eyes, he was a philosopher who would come to guide and challenge my views on the world. While Ronnie was far from perfect in a conventional sense, I firmly believe he gave me an extraordinary insight into the world and helped me break a cycle I was probably doomed to repeat had it not been for his influence. In the years during which my outcome could have been very different, Ronnie was the person who stepped up and said, 'Matt, push yourself, do better, go further and be more'. For this I will be eternally grateful.

In the years after I finished school and started to make my way in the world, which led my path to cross with Kaz's, there is one thing I learned, one skill I honed and continue to hone, and that is the ability to endure in the knowledge that life owes you nothing but the breath in your lungs and that you will be tested time and time again. Your will to endure shall be pounded and beaten into you like a blade under a blacksmith's hammer. You will be beaten and shaped and you will be tested. And if you endure the process, you will grow stronger. I have learned to endure and to keep on enduring despite the many tests thrown at me. I may not be the best at anything or the strongest, the bravest or the best looking, but I will endure.

As I began to navigate the tough journey of separation after yet another failed marriage, it was comforting to know I now had two people in my life who were going to be there for me unconditionally, without judgement – Jussy and Matt. Jussy and I had been best mates for about ten years. She had pretty much seen it all. She was my rock throughout my first mistake… oops, I meant first husband! But, in all honesty, marrying him really was a mistake. The only good that came out of that relationship was our daughter, Jess.

Jussy and I didn't have the best start to our friendship, but we soon realised that we really liked each other, despite our differences, and have always had each other's backs through thick and thin, always have, always will. We are similar and yet so different. Jussy is refined yet can rough it with the best when required, whereas I am the opposite. In fact, I'm the absolute opposite of refined. Rough around the edges, nope, rough to the bone! However, when the need arises, I can mind my P's and Q's too. I guess you could say we balance each other; I'm the yang to her yin.

Once I finally mustered the strength to call it quits on my first marriage, my ex happily went his way and I went mine, much to my father's disgust. My decision to leave my first husband caused a massive fight between my father and me. It was the last time he ever raised a hand to me, mainly because I was finally big enough to tell him I would knock him to the ground if he dared touch me again. There were only a few times in my life that my father hit me. I'm sure if you asked him, he would tell you I deserved it every time. My father was a hard man and when he was angry you knew it; he could freeze you out for months.

When I fell pregnant with Jess at nineteen and my father found out about the pregnancy, he called me, told me to get an abortion and said he would pay for it and we could put it behind us as though it had never happened. I refused, he hung up on me and we didn't talk for seven months. So here I was, pissing off my father yet again by failing in my first marriage. Due to all the fighting, Jess and I had to leave the granny flat we lived in on my parents' property and move in with Jussy. I'm pretty sure all of Hornsby thought we were lesbians. But we didn't have a care in the world about that; they could think all they wanted. We lived together for a while, Jussy, Jess and I. Jess was so little and had no clue what was going on. She suddenly had two mums who loved her and made her life fun. I was doing shift work at Westpac and Jussy did parental night shift for me, looking after Jess. Jussy was a godsend, a true lifesaver. Without her, I would never have made it through that period. I wouldn't have been able to keep my job with the bank and I wouldn't have been able to move forward. She really was the person that picked me up through the first, and even to a degree, the second divorce. For the past twenty-five plus years, Jussy and I have continued to be in each other's lives. We are 'aunties' to each other's kids, we are 'sisters' and we know we will be there for each other whenever the need arises. She has been and will continue to be an integral part of my life.

Here I was, twenty-eight with two young daughters and navigating my way through the end of my second marriage. Was it fate that Matt came into my life when he did? I really don't know. All I know is that we just clicked from the get go. It wasn't long before Matt picked up on the fact that my marriage was in the toilet. He became an ear, let me talk it through and put things in perspective. He would listen, offer opinions and just encourage me to go for whatever it was I really wanted in my life, to follow my heart. As it turns out, he was having issues in his relationship as well. In fact, he was pretty sure his girlfriend had feelings for his best friend and that something was going on between them. So I became his sounding board as well, not that I had much advice to give at that stage, especially not with two shitty marriages under the belt. We listened to each other without judgement. Whatever decision either of us chose to go, the other was going to support.

I got the strength to finally call it with my ex. Not long after my relationship ended, so did Matt's after he confirmed what he suspected was happening between his girlfriend and his best friend. Matt and I became closer, neither one of us realising just how dependent we were becoming on each other. As my world seemed to be collapsing for a second time, this time with two children reliant on me, I had no choice but to rebuild all over again. Only I was luckier this time because I not only had Jussy in my corner, I had Matt too. He became the voice of reason that kept telling me the kids would be fine; the person who reassured me that I wasn't a complete failure as a mum; the person who held me up when I just wanted to collapse; and the person who saw the strength in me and encouraged me to draw on it when I needed it most. He was so young, yet he seemed to just get it. He was awesome with my girls too. Jess and Ayleigh took to him like ducks to water and he had so much time and patience for them. I can't explain what it feels like for a child to go through a divorce, but I do know what it's like to feel guilt as a parent for leaving your child's father. But sometimes sitting and just watching Matt with the girls would melt all that guilt away. They seemed to feel safe with him, protected almost. The more time he spent with them, the more the clouds just seemed to lift for them.

I often say Matt has an old soul; something I believe has a lot to do with his upbringing. He would use books as an escape growing up; he read a lot and thrived on learning. He also had older people, such as his Pa and Ronnie, as positive influences in his life. They made him think and see things well beyond his years. He would often talk about topics I had no clue about, but it seemed he felt safe enough to talk about them with me so I just listened. He had a very strong recall capability and could talk with just about anyone about virtually any topic. He would discuss Greek mythology one moment and then seamlessly transition to an in-depth discussion on science and politics. This ensured he fitted in well in any circle. He was confident meeting new people and respectful of people as well, which has always been very important to me. We were good mates and he was helping me get through a really difficult time. And I think I was helping him too, as he was struggling at home and still not over his breakup.

MATT'S PERSPECTIVE...

We seemed to just get along really well. I liked that it was such an easy, natural bond. There wasn't anything clandestine about it; I just saw Kaz as a mate. She was my friend and I could see that she was hurting, and I wanted to help, I wanted to support her, I wanted to be there for her. The more time I spent with Kaz, the more I learnt about her. She was like a diamond in the rough; there was an inherent value there that not everyone could see. She was fun, she was easy to be around, and she was strong, even if she didn't realise it at the time.

I was in a place where I felt I had no one other than Kaz I could really talk to or unload anything onto. Home certainly wasn't a place for me to go to and talk to anyone. And when your girlfriend hooks up with your best mate, the friend network you thought you had immediately goes into shutdown mode. It was priceless for me to have a place to go and a person I could turn to and trust, knowing I wouldn't be judged and that I could be completely confident that anything I said wouldn't be used against me or become fodder for the gossip mill. I guess at the end of the day, we just gave each other the strength to get through a really shitty time.

CHAPTER 2

The rumours become the truth

As the friendship between Matt and I grew stronger, days turned into weeks and we seemed to heal from all the other crap going on around us. We became each other's voice of reason and support. We started to become inseparable. Matt was at my place almost every other day. He'd go to work, come to mine, hang out, have dinner, help with the kids, go home and repeat it all over again. Every other day soon became every day and eventually it just made sense. Matt moved into my spare room. I needed help with the bills, and he wanted to get out of his parents' house. And we were hanging out together every night anyway. It just made sense, well to us anyway.

But, as our friendship grew, so did the rumours and innuendo. I remember one time when I was out shopping with a mutual friend and the checkout chick at Target, who knew my friend but had no idea who I was, asked her if she knew what Pagey was up to these days. My friend knew exactly where the conversation was going but she played along.

'No, what about Pagey?' asked Kate, feigning curiosity and surprise, while trying not to look at me.

Looking incredibly pleased with herself, the checkout chick leaned forward and said, 'I heard he's having an affair with an older, married woman – much older. She lives up at Kariong. Apparently, Pagey broke up the marriage, and now he's moved into this woman's house'.

Kate, bless her cotton socks, was never one to miss an opportunity to take the piss out of someone. She could not have played it any better.

With a massive grin on her face, she simply said to this unsuspecting chick, 'Hey, this is my friend, Karen. She's the woman Pagey is living with in Kariong'.

You should have seen the look on that poor girl's face. I nearly lost it; I was trying so hard not to burst out laughing. She had no idea what to say.

To break the awkward silence, I looked her squarely in the eye and said, 'Yes, Matt has moved into my house as a boarder. Yes, I am getting a divorce, and no, I am not sleeping with Matt. But we are really good friends. Have a nice day. It was lovely to meet you'.

And with that, I held my head high and walked out of the store with Kate. I still laugh when I think about the look on the poor girl's face. And it was all true, well at that stage anyway.

Matt had turned nineteen by the time he moved into my spare room and I was twenty-nine. His parents were not happy about it and neither were mine. Our close friends knew we were just really good friends, although some of them, especially Jussy, have since said the writing was always on the wall. Those that weren't close to us simply assumed we were sleeping together and said, behind our backs, that it would never last. How wrong they were, on both counts.

I guess when we look back at it objectively, we were getting much closer than just friends, especially once Matt moved into the house. We made each other laugh and I looked forward to coming home from work and seeing him and the kids. Home became a much happier place. We would still sit up late most nights and talk about nothing and everything; we would work out, cook meals and clean the house together, all the things a happy couple would do. We just didn't realise it.

Then one night everything changed. It was pretty late, and we were both in the office. Recently converted from my old garage, the office had the bare essentials: a desk and carpet and a couch. I'm not even sure how it happened. We were sitting together on the couch, talking, just as we'd done many times before. If you ask Matt he would say I kissed him. But I reckon he kissed me. In any case one of us kissed the other and the other didn't pull away. So a kiss became two and then three, and then one thing led to another and... well... you know how this goes.

The next day things got a little awkward. Neither of us was really sure what this meant and neither of us wanted the world to know. We both agreed to be friends with benefits and not tell a soul, well maybe just Jussy.

We were like that for a few months, but deep down we both knew we were no longer just friends with benefits. We had developed strong feelings for one another. And then things became really serious. I fell pregnant and miscarried before I even realised I was pregnant! It was all kinds of messed up. Matt and I had been playing it safe and the thought of having any more children hadn't entered my mind. That option had been taken off the table when my ex got himself fixed up, so somehow it never occurred to me that this might happen. Looking back now I know how naïve I was. But then all of a sudden, I developed a sense of wanting it too. And I felt an indescribable sadness that the decision had been taken from me. Or maybe I really didn't want any more kids and losing the baby was a blessing. I didn't even know I was pregnant so why was I feeling this way? I needed time to think and make sense of it all. I had no idea how to feel about it because, in the end, none of it mattered anyway; the decision was taken away from me. Loss... that's all I felt... an aching loss over something I never knew I had until it was gone.

I was too afraid to tell Matt. I wasn't sure how or what he would say and what this would mean for us. I was only just starting to get things back on track with my life after a shitty settlement with my ex. I didn't need another blow. Right, wrong or indifferent, I shut down and went into self-protection mode. It's what I do. The only person I told was Jussy.

Matt had never seen me so closed off before. It hit him hard and he had no idea how to handle it. Jussy made me realise that I had to sit down with Matt and talk to him, explain what had happened and give him the chance to process it too. But my shutdown had triggered something in Matt. It turns out he was going through some internal struggles of his own, struggles that caused doubt to enter his mind about our relationship. So before I had the chance to tell him about the pregnancy and subsequent loss, he told me he couldn't be a part of this relationship anymore, that the friendship between us was really important to him but a long-term relationship with me was not something he could commit to. He didn't actually say why at that point, just that he needed out, or maybe he did and I didn't want to hear it. I was a bit shell-shocked at this point. I went into panic mode. I was losing my best friend and I didn't know why or what to do or how to stop everything from snowballing out of control. So much loss, all in one hit.

As always, Jussy was there by my side, knowing all, seeing all and supporting how and where she could. I still had to get up every day and put my best foot forward. None of this was Jess or Ayleigh's problems and those kids had been through enough already. So I was going to keep smiling, no matter how much I felt like I was dying on the inside. For some reason Matt was still living in my house and it broke my heart to see him there but not really present. I was so scared of losing one of the best people I had ever met. And then on top of everything else, he found out about the miscarriage. Through it all, there was still something so strong between us and he was just there, there as my rock, helping me get through it all when he should have run. He was young and had no idea about life and it was messy and complicated. Most people his age would have run as fast as they could, but not Matt. I have no idea what made him stay.

MATT'S PERSPECTIVE...

The last thing I was looking for as a nineteen-year-old kid was a ten-year-my-senior divorcee and a ready-made family. If someone had told me the year before that this is where life would take me I would not have believed it. But that's life, I guess. Kaz and I are both dreamers, and the trouble with dreamers is that we can't be told that something is impossible. We live the impossible. It's never been a thing to deter us. However, looking back now, I do believe Kaz was more open to the possibility of what we could become than I was in the early days. I'll be honest; the idea of raising two kids that weren't my own and accepting that I might never have my own kids was well beyond my ability to deal with. But Kaz reassured me that I was already a great influence on her kids and that we would have our own kids one day. It's just that the path towards getting there was a little rocky. During our journey from friends to a full-blown couple we had many hurdles to overcome. I'll admit to having had second thoughts about the relationship and, life being what it was at the time, I was unaware that Kaz had suffered a miscarriage. When these two events collided it raised a real whirlwind of thoughts and feelings in me that I had no idea how to manage. And so, like most people, I tried to convince myself that I was doing the right thing by squashing the relationship. But our bond proved to be too strong and we worked our way back. Don't get me wrong, this is no fairytale and our life isn't perfect. But for whatever reason, we just seem to work.

I wasn't opposed to having another child, but the timing wasn't right. Feelings of fear and loss threatened to consume me to the point where all I wanted to do was run and hide. But I couldn't. Jess and Ayleigh needed a mum who was strong and could get up every day and make the world okay for them: keep house, pay bills, go to work, feed and clothe them and not look like a blubbering mess every time they saw me. Smile, bury the sadness and keep going. I'm pretty sure Matt was confused as well. He didn't move out and I didn't ask him to. He still had this need to protect us and I'm pretty sure he also wanted to be a part of something, maybe at this point just a part of the healing, maybe more. Neither of us really knew what that 'something' was.

After weeks of living like this, I knew it was time to open up the lines of communication and confront what had happened. I needed to know where it left us. So I arranged a night for just the two of us, with no one around, no interruptions; a night where we could just put all our cards on the table, work out where to from here and respect each other's decision. We started to talk about what would have been when all of a sudden he opened up about what was really going on in his mind. He explained that he had freaked out because he wanted kids of his own one day, and, as I already had the two girls, he assumed that I probably didn't want any more kids. Then when he found out what happened to me and I didn't turn to him straightaway, he was hurt. I got that. I should have let him in; I should have had faith in him that he wouldn't run, that he wouldn't blame me. Matt was different from anyone I had ever been with. I told him I had no issues with being a mum again, and that if we were meant to be, kids would absolutely be part of the journey for us. I made him a promise that night. And it's a promise I had every intention of keeping. By the end of the night we both reached the same conclusion: that we weren't done. So we decided to go slow and let the cards fall where they may.

As one day rolled into the next, we rebuilt from where we had left off, both agreeing we would stay in the closet until the time was right to tell the world. Not even the girls knew. Only Jussy knew the truth. It was such a surreal time. It was almost like I was a little kid with a big secret, a happy secret that made me feel on top of the world. I was in such a great place with almost every aspect of my life for the first time in a very long time. I was loving my work and revelling in the added responsibilities I had taken on in my professional role.

Work has always been such a huge part of my life. I know I am smart, but the truth is, I never actually learnt how to learn, not in an academic sense anyway. I was a troubled student, always accused of being disruptive and non-compliant. They would write on every report, 'Karen has the ability, not the initiative'. I wanted to learn, I just didn't know how to; I didn't know how to retain the information they were telling us. I'm sure there are many labels that would be given to me today, but back then, labels like ADHD, ADD, learning disability and so forth, didn't exist. As far as my parents and the teachers were concerned, I was just lazy and a bit of a clown. I guess that's why achieving in my work was and continues to be my way of proving to myself that they were all wrong. I love achieving. I love bringing things to life and seeing the benefits. At the time, I was working as a systems administrator for an amazing company called CHEP. I started with them as a team leader for the software help desk. Before joining them I had spent ten years working for other companies within the IT sector, learning everything I knew about computers through hands-on experience. Matt was working in real estate at the time but he didn't really know what he wanted to do long-term. He was young and still figuring out who he was.

It was wonderful watching my kids find solid ground again and feel comfortable and stable and happy. They loved having 'Uncle Matt', as they called him, live with us. Matt didn't seem at all perturbed by this. In fact, he seemed to revel in it. He had so much patience where my girls were concerned. He would read bedtime stories and help them with their hair and make sure they had done things like brush their teeth before bed. He was so much more than a flatmate, not only to me, but to them as well.

Matt's 20th birthday was rapidly approaching and we had already started talking about the fact that we couldn't stay in the closet forever. In the space of just five months we had started a relationship, broken up, rekindled the relationship and made the decision to give it a serious go, all under the cloak of secrecy. Matt's parents were slowly getting used to the fact that he lived with me now. His mum and I were getting along really well and she was lovely to the kids. But his dad kept his distance. While I'm sure he wasn't overly happy about the situation, he was polite and friendly enough when we came together for any reason. My father, on the other hand, was a very different story. As I said, Dad was a hard man and more stubborn than an ox. He was totally opposed to Matt and I living together and made no secret of it. He was cold and cranky on the few occasions that he would speak to me. But it was all getting a little too difficult to keep the secret going. And my divorce and the fight my ex seemed determined to have with me was starting to make life pretty hard financially as well. It was becoming increasingly clear that he was going for everything, including the house the girls, Matt and I were living in. We needed to make some big decisions, including where we would live, and we needed to make them soon. So, for the very first time, we started talking about our future, what it might look like and where we might live.

Matt and I decided that first things first; we needed to let our friends and family know we were together. It was time to come out of the closet. It turned out to be quite an experience and not the way I had intended for us to do it, but that is Matt to the core. His mum and I decided to throw him a surprise 20th birthday party. Well, as it turned out, the only surprise on the night was Matt's speech. Only two people at the party beyond Matt and I knew our secret; well that was the case until he announced it in the last part of his speech!

It was pretty blunt too. He had all twenty plus people listening, a captivated audience. He thanked his Mum and I for a great evening, thanked those who travelled and everyone else for coming etc. etc. and then, just as I thought he was wrapping up the speech, he said, almost as an afterthought, 'So thanks again everyone for coming and there's just one last thing I'd like to say: Kaz and I are together as a couple, so cheers and have a great night'. Holy shit, you could have heard a pin drop! The response was mixed to say the least. Seventeen years later and I still giggle about that night.

So it was official. We were finally out in the open. I must say it was nice and it felt like the clouds were finally lifting. But there was still the matter of our living situation to figure out. By now it was obvious that I was going to lose the house to my ex. Things were pretty bad between us and he was going to do whatever he could to hurt me and come out on top financially. Matt and I needed to figure out where we would go, whether or not we would stay on the Central Coast and deal with the naysayers or move away from our network and start afresh. The more I thought about it, the more I started to search further afield. If we were going to move, where would we go? Where could we afford to go? I started researching online. I was looking for two factors. Where could we get work? And where could we afford to buy our own place? As I had always wanted acreage with a couple of horses, and there was no way we could afford to buy acreage on the Central Coast, we decided it was time to go and make a new start with a clean slate.

CHAPTER 3

Reality bites

I was at work one day having a scan around the Internet when I stumbled upon a university that was offering work I could do. It was just admin work, but it was a start, and it would pay the bills. I looked at the real estate there and even small acreages seemed affordable. I called Matt and asked, 'Hey, wanna move to Armidale?' 'Yeah sure, where's Armidale?' he replied. 'I don't know. I'll ring Mum and ask,' I laughed.

Within weeks we decided we would move to the rural New South Wales town of Armidale. We could afford land there, I could get work at the University and everything would be great. Matt still wasn't sure what he wanted to do long-term, but as a real estate agent, he could pretty much work anywhere. So off to Armidale we went, a town neither of us knew anything about and a place where nobody knew us.

MATT'S PERSPECTIVE...

I had nothing really keeping me on the Central Coast, so when Kaz called me one day and asked if wanted to move to Armidale, well what could I say but, yes! I knew that owning her own patch of land was a dream for Kaz and if the only thing standing in her way was me, well, who was I to stand in her way? So, true to our natures at the time, we put it together and I moved up in the November to get a house sorted out. Kaz followed several weeks later with the kids after wrapping up loose ends. As far as emotions go I tend to be fairly pragmatic. If I find myself in a new environment or set of circumstances I tend to treat it as my new normal. So the decision to make the move and the subsequent settling in process was just my new normal. The tough part was the everyday stuff: working out the way a small town works, attempting to ply a trade you knew in a completely foreign environment and the trials of a young man who still didn't fully understand who he was or what he wanted to do with his life. I secured a job straightaway, working for a real estate agency in town, but I struggled in this role because the man I worked with was unscrupulous and underhanded. Not only did I struggle with the manner in which he conducted business, but also with the things he expected of me. It left me jaded and resentful. In the end I started preparing to leave, which gave him all the ammunition he wanted to give me the boot about seven months later, not all that long after we bought the farm.

Before considering the job at the Uni I had discussed the option of staying with CHEP and working remotely. Unfortunately, the manager at the time didn't quite understand the full concept of tele-commuting and felt it necessary for me to travel back to Sydney every second week. That was never going to work. So, I decided to apply for work at the University of New England. I met a woman by the name of Valerie Dalton who would soon become not only my new manager but also a very strong support person for me in a battle I did not see coming. I quit my job with CHEP and was lined up to start my new job at the University and, by Christmas 2002, we had settled into life in Armidale.

Armidale was a good five hours away from family and friends back home on the Central Coast. The girls started at the local school, made new friends and started afresh, just as Matt and I got to as well. It also meant that the only father the girls really knew – my second husband – was also now five hours away. I'm sure he hated the fact that we moved so far away, however as we stayed within the state, there wasn't a lot he could do about it. He made it so hard for me to stay financially, so I didn't feel bad for moving. I wanted my kids to have a comfortable life and I couldn't afford to do that where I was. We made arrangements that we would alternate school holidays and Christmas etc. and he drove to Armidale about every third weekend and stayed in a motel with the kids for the weekend.

We rented a little two-bedroom place and within six months we had purchased a tin shed on a seven-acre property in a little place called Invergowrie, about fifteen minutes out of town. We had made a few friends by this stage and all of them had the same thing to say when we told them where we had bought: 'You bought all the way out there?' they would exclaim in surprise, as if it was going to take them days to come visit us. I thought it was quite amusing, as you wouldn't think twice about moving that distance on the Central Coast. In any case we bought it, we could afford it and now we had our work cut out for us to make a comfortable home for the four of us out of the tiny one-bedroom converted tin shed.

By this time Matt had work with a local real estate agency and I had my job with the University. Over the next twelve months we spent almost all our money and spare time renovating, trying to get it to the point where it was more of a home than a shed. It would forever be the tin shed to us though. There were actually two sheds side-by-side on the property: one was completely empty and the other consisted of our small bedroom, kitchen and bathroom. It brought a whole new meaning to the word 'rustic'. My God it was rough! But it was ours and we were improving it little by little. We even learnt how to insulate plumbing so our pipes didn't freeze in winter. Eventually we turned the first shed into a cosy little two-bedroom home. During that period, it really started to feel like Armidale was home. We had horses and dogs and we were making friends. Some were friends for a period, others were not really friends at all, and we eventually had to find that out the hard way. And some are still and will always be friends. It was a great period because no one knew us or judged us. We were just a family new to town making a go of it. We grew as a couple and as a family. Matt stepped up as a father figure to Jess and Ayleigh; he didn't miss a beat with those girls. Sometimes they fought against his requests, other times they just went with it. With each passing day they looked up to him more and more. Even though they still called him Uncle Matt, he was Dad now. While we had our challenges turning our little farm and shed into a home, we also had a lot of great times during those twelve months. Little did I know that dark times were just around the corner.

Matt's career in real estate never took off in Armidale. His boss had questionable morals at best and not a great reputation in town. So Matt never settled there and, in the end, they parted ways. At about twelve months into my new role as an administrator at the University, things just seemed to get heavy in my head. I couldn't explain it other than a sense of constant sadness. I felt like I wanted to give up, but I had no idea what I wanted to give up on, or why. It became all-consuming; there was no warning when or why it would hit. It was like a black cloud rolling in just before a dump of rain; only instead of rain it was a massive dump of tears over which I had no control. It made me angry because I hate to cry, which made me lash out at the ones closest to me because they wanted me to explain what was wrong and I couldn't; I had no idea. It was affecting me at home, at work, even in transit between the two. It was changing me, and not for the good. I couldn't let anyone in and I felt like I couldn't get help because I didn't know what was wrong. I just couldn't lift the cloud. It got so bad that I would find myself losing it, crying uncontrollably for no reason whatsoever. I didn't want to see anyone at that point. I wanted to hide; I wanted to cry in silence with the world shut out. Each day seemed to take forever to end and, in my mind, a new day wasn't a blessing, it was a curse. I hated my life and that hurt. It hurt because my life was pretty damn good at this point. I should have been grateful and happy, not losing it day in, day out. I had a man who loved me unconditionally and supported me on every level. He made me laugh and made me feel safe. I had two daughters I adored, each so different from the other. It was amazing watching them grow and find their own uniqueness. I had a home, land, dogs and horses. Everything I had ever wanted. Why wasn't it enough? Why couldn't I just be happy? Why was I feeling like the world was falling in on me? I couldn't explain it and I was getting worse by the day.

I was one of the lucky ones. Two very significant people in my life pulled me through some of the darkest days of my life, Matt and my boss at the Uni, Valerie. Matt was my rock, and that was never in question. But it was Valerie who saw the signs and recognised depression. To this day I still have no idea why it hit me. Maybe it was just a combination of a shitty childhood and a shitty early adulthood, most of the latter courtesy of my own bad choices. Valerie was amazing. More than a boss, she was also my friend. We connected from day one. I remember one day she just gently started talking to me. She would take me for walks through the University gardens and listen to a whole lot of rambling from me, none of which probably made any sense at all, but Valerie just listened. Every so often and at what seemed like just the right time, she would offer a suggestion or relate my experience to something similar in her world. At the time it kept us connected. Unbeknown to me, Valerie already knew what I was heading into and the help I needed to get through. But she also knew me well enough by then to know that telling me straight up that I had depression and needed medication and professional help was never going to fly. That would have been like waving a red flag to a bull. I was raised in a world where depression was a bad word used by weak people hiding from their responsibilities in life. Not real. I think even in today's society, there are far too many people who either don't or won't acknowledge depression, anxiety and other mental health related conditions. To the other extreme, I also believe we have made it so easy these days to cry wolf that many do. It's such a hard space, but I just know we need to do more, research more, understand more and be more open to finding cures, not Band-Aids.

Slowly and carefully, Valerie coaxed me in the right direction to seek help. With Matt backing me every step of the way, I finally took myself to see the people that could help. And between doctors, psychologists, Valerie and Matt, they all helped me get to a point where I began to get my strength back. Well, them and the little white happy pill. Ah, the little white pill! I talk about this a lot with people. To this day I still hear people sound just like I did all those years ago. 'I'm not taking anti-depressants, you are not going to chemically rearrange my brain!' What I learnt was that my brain had a chemical imbalance. I had endured a lot and had never really been taught how to deal with the emotional upheavals I had been through, either at my own hands or at the hands of others. My coping mechanism was to bury and ignore, to soldier on. It eventually came back and bit me on the arse.

I also learned that if you are at the point where the fog just never wants to lift and the weight of it hurts every day, like an indescribable pain, and the only way to get respite, the only way to get enough of a break so that you can breathe and stand even a slim chance of rebuilding yourself is to take a little white pill every day, then take the pill. Ignore the stigma and take the fucking pill.

I took that happy pill for two years. Throughout this period, this little white pill gave me time to gain clarity; it allowed me the time I needed to sort out the demons in my head, to deal with what was and should have been; to separate the demon that is depression from reality. It didn't stop the depression, but it did help me gain enough stability to see it, acknowledge it and work towards finding the root cause. I'm not sure what the alternative would have looked like had it not been for the support and guidance of Valerie and Matt. I probably wouldn't have got through. Matt took the stigma away for me. He was the one who helped me reconcile taking the pill. He was the one who sat there with me till the early hours of the morning, listening to me go back and forth about why I shouldn't take the damn pill and why I should. He helped me work through what I thought was weakness, the shame of not being strong enough to fix it on my own. It was our secret and only those that had to know, knew. Matt never made a scene or a deal of it; he just quietly and constantly made sure I was okay.

MATT'S PERSPECTIVE...

Most people would know what it's like to have someone you love in pain and not know how to fix it for them. And that's exactly where I was. Logically, I knew that everything was cool and that our world wasn't falling apart and that better times would come. But no amount of knowing that will make it okay for the person who's deep down in the hole. You just have to put in the work, both of you. And that could mean taking medication, getting professional help, venting your spleen, going for a walk or all of those things. The thing with depression is that the person suffering with it has no ability to do all those things completely for themselves in the beginning. These are all skills that you learn through the work of living with this burden we call depression. For Kaz, it was no different. I wanted to fix everything but no amount of fixing anything helps at this point. The cycle has to work itself through and the best you can do is support, be there, listen and help them work through the cycle. For Kaz and me, sometimes we worked through it together, other times I had to drag her through, kicking and screaming.

I had to reassess everything, including, and especially, myself. What I wanted for me, what I could live with and what I couldn't. I had to make some tough decisions. And one of them concerned my career. I made the decision that I was not okay with being an administrator in a University. So I reconnected with CHEP. The lesson from this is never ever burn bridges. You just never know when you may want or need to cross them again. After an eighteen-month sabbatical – well, that's what I'm calling it anyway – I got my old job back. The role had expanded since I left, meaning I was now responsible for the entire Asia Pacific region, eight countries in total. It opened the door to exciting new opportunities for me and really boosted my confidence and self-esteem. I finally realised that one of the things I'd been missing was having a career, not just a job. And I wanted it for me, not because of what others may think of me or to satisfy my ego. From that moment on I let go of other people's opinions. I still care what people think, just not too much, especially if their opinion goes against what I want and what is right for my family. Call me spoilt, petulant or argumentative, I've heard it all before and it's true. I'm a fighter. I will fight for the underdog. I will fight for what is right. And at that point in my life, I was fighting for me. So in July 2003 I parted ways with the University, but not with Valerie and her family. They became lifelong friends. I will forever be indebted to her for the support she gave me when I needed it most.

In my new and improved role with CHEP, it didn't matter where I was located as long as I had a computer, a phone and I was near an airport. It gave me the opportunity to expand my knowledge in business and systems processes and design as well as teach others. The management team was very supportive and for the next ten years I got to travel to a lot of countries and meet and work with some amazing people; some are still very much in our lives as part of our family today. There is always a line for me between home and work and, to a certain extent, I am a very different person within each role. I try not to blur the lines; friends are friends, colleagues are colleagues and really good friends are family. Sometimes though, I have been lucky that the lines crossed. Again, lesson learnt, never draw a line in cement; you can't move it. Be open to opportunity, even if it sometimes goes against a rule you created yourself.

The little white pill also gave me the clarity and strength, albeit with the addition of a few too many beers, to finally confront my parents. It was Christmas Eve, 2003, the year I turned thirty-one. The massive showdown started between Mum and I, then Dad got involved. Lesson learnt from that night: don't drink alcohol with my parents ever again. We never spoke of that night again either. I can laugh now only because I can still picture my father's face, that raw look of horror and shock when I told him a few things he didn't know, one of them being about the night his daughter was raped at fourteen and never had the courage to tell him or anyone, and that this was why I was walking the street and not in the caravan with my friend when they found me. I remember him falling back onto the stove in the kitchen as the realisation hit him like a sledgehammer that his only daughter had been through things no one should ever have to go through and the knowledge that I endured the protracted trauma of that event on my own because neither he nor my mother had ever taught me that it was okay to lean on them. My relationship with my dad changed that night. From that night on he had no choice but to see me differently. He knew, no matter what, that I'd make it through. With or without his support, blessing or understanding, I would make my own way. I learnt how to make my life right no matter the journey. I was stronger than anyone in my family because I had endured more than anyone in my family all on my own. And I was still standing, still smiling and still willing to give it my all without asking for anyone's help. I wouldn't hide behind alcohol or gambling as others do; I wouldn't live my life through superficial pleasures and forsake being real for a life of plastic. He finally knew who I was and I was finally able to stop needing him to know, or even needing him to approve.

I don't fully understand what that confrontation did to my parents that night. What I do know is that it wasn't about them or for them; that night was for me. That night I got to say it, I got to stop being the naughty child, the non-achiever, the bad kid. That night I got to tell them they fucked up as parents. That night I got to let go and I did it because the one person that truly believed in me and loved me unconditionally was there to hold me up. I had Matt by my side, and I finally got to clear the demons. I never took shit from my father again.

CHAPTER 4
The Green Machine

Life was rolling along nicely when Matt made a decision that was to alter the course of our future; we just didn't know to what degree at that point. It was a day like any other when Matt came home and told me he wanted to join the Defence Force. It didn't come as a real shock; he had talked about it on and off for a while. But this was the first time he was really serious about it, and so we had to start talking about what that would mean for us as a family.

MATT'S PERSPECTIVE...

It was tough being the new kids on the block in a small country town trying to fit in and find steady work. Nearing the end of our renovation, I commenced the process of enlistment into the Australian Defence Force. My family had always had a sporadic relationship with the ADF, AIF and as Commonwealth soldiers. So it was on the cards as potentially something I could do after leaving school. I just wasn't ready for it at the time. But I was looking for a job now, so why not? I knew the process to get into the role I was looking to do within the Army was going to take a while, so I was relieved to be able to pick up some contract work at the University in the meantime. It wasn't reliable work though and I never knew from contract to contract what, if any, work would come next. So here I was, looking down the barrel of yet another period of unemployment. Well, fuck that! I could not and would not let myself be put in a situation where I was not contributing financially to the household. So I walked out of the office I was in and went straight to the local Army Reserve Unit to speak with their recruiting staff about how to speed up the process of enrolment. The long and the short of it is that, after much discussion with Kaz and going back and forth with the decision, I bit the bullet and decided to join the Army Reserves, as it would fill the gap between contract work and ensure I was contributing, while also allowing me to transfer should circumstances change. As it turned out, I made the right decision, as it allowed me to keep my relationship with my future wife and gave me an opportunity to do the job I wanted to do, as well as stay in the black financially.

I always tell people I married the man who became a soldier. And that was the truth. Even though we didn't officially tie the knot until much later, we were married for all intents and purposes. We talked about the implications of Matt joining the Army full-time. It was early 2004 when he began enquiring into full-time – or regular, as they call it – Army. With two small children I was scared, uncertain and unwilling to be constantly moving them around for postings every few years. I'm not sure if, at that moment, I was being completely fair to Matt. Perhaps I should have just sucked it up and gone with him on the idea of going full-time. Perhaps everything would have been okay. Other families survive and kids are so adaptable. It's hindsight now and I will never really know the answer, but my fear of the unknown won and so, as a compromise, Matt and I agreed that he would join as a reservist. If he liked it and was still committed to joining full time after a few years, then I would back it. How naïve I was!

So, Matt joined the local Army Reserve Unit, officially known as the 12th/16th Light Horse Brigade, Armidale. He would go in once or twice a week in full uniform to do 'stuff'. He really seemed to enjoy it and took great pride in getting his uniform ready. After a session he would always come home inspired, full of information about the people, their careers and how it all worked. I knew this was the next chapter in his life and it seemed he was finally finding his feet.

I personally think he should have gone to University and become a scientist or a medical doctor, as he was like a sponge; he could learn anything and everything he put his mind to. However it seems the Australian Army was to be his calling.

Matt had to go through basic training, but that wouldn't commence until early 2005. So for the next eight months or so he became part of the 'Chocs', a derogatory term given by full-time soldiers to part-time soldiers that had since stuck and become the norm. For all intents and purposes Matt was now a Choc. While he wasn't really doing much other than turning up for parade nights and learning the basics within the barracks, he was very quickly becoming one of the boys and loving it. During this time, he was still working on contract with the University, but you could tell his interest, his spark, was completely embedded in the Green Machine, the Australian Defence Force. He finally seemed to be finding his place in life beyond the kids and me. It was a happy period in our lives. We were in our little shed on the farm fifteen minutes out of town, we were slowly building it into a home, we had some horses, the kids attended a good school, we were making some really strong friendships, I was back at CHEP in a role I loved, and Matt was finding his passion. I was building a really strong career for myself, so I got it, I got the importance of needing to fill that void, of wanting to be a part of something bigger and feel that getting up every day and going to work needs to be for more than just a job. Life for us was moving forward and we were enjoying it.

In early 2005 Matt went away for six weeks' basic training as a Choc. Minimal commitment they said. How naïve I was! It was the first time Matt had ever really been away from us since being together. I would travel for work and that was normal, but now it was Matt's turn. He was a great person, an awesome partner and an amazing step-dad, but he was still very young and had no real clue about life yet. General day-to-day tasks like washing and tidying – you know all that boring stuff – were usually happily ignored by Matt. It was a small price to pay. I didn't expect much of that to change really, as I had no idea what would happen while he was away. I guess I really didn't expect anything to change. I also didn't really expect to hear from him. We had been told a few very different stories about what to expect so I figured it was best to assume nothing.

After being away for just a few days I was surprised to get a call from Matt. We chatted for a bit. I told him how great it was to hear his voice and he told me the same. But I could tell something wasn't right. It took all my strength to hold back and wait for him to tell me what was on his mind. Continuing the small talk, Matt seemed all over the shop when suddenly he blurted out that he felt he had made a mistake and that he wanted to pull the plug and come home. I was shocked. I wasn't really sure what to say, so I just listened for a bit. He told me all kinds of reasons for feeling this way, however none of them really seemed strong enough to warrant pulling the pin, especially not on something he had wanted so badly. I told him I would back whatever decision he made but implored him to take a few more days to think it over before making a final decision.

I remember my response on the phone that day as though it was yesterday. 'I think it's all just so new and a bit of a shock. But I know you are good enough; I know you can do it and you will get through this, Matt. And I honestly think you will regret it if you walk away now.' I often wonder how different life would have been had I just said, 'Come home'.

MATT'S PERSPECTIVE...

I was enlisted late 2004 and attended the Army Recruit Training Centre, aka Kapooka, for basic training. The rest of the time was spent working, getting ready to go to basic training and putting a plan in place for how it was all going to work. It was probably the second or third night in. We were marched to a small canteen, like the kind you might find at a caravan park in a remote township. It had all the necessities: toiletries, stationery, batteries and the like. For the lucky few who were allowed to consume them, it also had an assortment of confectionery, soft drinks, biscuits and ice creams, however none of the latter were for us; we hadn't earned it. In any case we weren't there for supplies from the canteen. We were there to see the barber next door for our one-size-fits-all haircut: a number three all over, which was a waste of time for me as I wore my hair cropped as close to the skin as possible. Nevertheless, one in, all in. So, like everyone else, I had to wait in line, step up when called, take a seat, have the barber run useless clippers over my head and pay for the pleasure of it all.

We also had an opportunity to make a phone call to our next of kin that day. I was away from what had become my life, away from my family for the first time ever. I had been away for a weekend here and there but, given I had essentially moved out of my childhood home to start a ready-made family with Kaz, I had always had family around me to weather the good times and the bad. Now I was staring down the barrel of six weeks away from my people, relearning how to be a human the way the Army wanted me to. And it rocked me. I have always been one of those people prone to melancholy, and would often beat myself up emotionally if I failed at something or felt I needed to motivate myself to do something about it. So being taken away from my family, and my comfort zone, and entrusted to the care of the 'tyrannical guardianship' of our section commander was challenging for me, to say the least. I can't remember everything Kaz and I spoke about when I called home that day, but I do remember telling her, 'I don't think I can do this'. My self-doubt was in overdrive and I couldn't see through the fog to know what was real and what was doubt, insecurity and just plain bullshit. Kaz pulled it around for me. She scooped up all the fractured mess that was my psyche, put it all back inside and reassured me that I could do it, that I had the ability and the strength to do this. Hearing her voice grounded me again, gave me perspective and returned me to reality. She always does. She is really the only person, even today, who can pull me out of the depths of despair and self-sabotage.

I made it through those forty-five days of basic training. And I was hooked. From that phone call onward I loved every moment of it and wanted to do as much work as possible. However, the position I was in meant that if I was to go full time, I would have to go back to Kapooka, wait for a spot to open up at a platoon and complete further training. The alternative was to go home, back to my reserve unit, and finish my training there while Kaz and I worked out all the details of what life would look like for us. In the meantime, I would pull as many work days as I could at the reserve depot and get as many courses in as possible while I was there. This all led to me slowly working more and more for the ADF and less and less for the University.

Matt's march-out parade at Kapooka to celebrate the completion of his training was an incredible day. His parents, one of his old schoolmates and I attended, and I can honestly say I was so proud of him that day. Matt loved every moment. He beamed from ear to ear. Everything was right with the world. He had finally found his feet and he wasn't letting go. He knew he would go full time, he just had to get me there as well, bide his time, wait for the right job, the right location or the right situation to come up and jump on it. Getting me on side with his decision to go full-time Army was now his mission. Anyone who knew him could clearly see the change in him. He was in, hook, line and sinker. In hindsight I guess I knew it was inevitable that he would go full time eventually and that when he did, we would be right there beside him, just not today. What I didn't realise in that moment is just how quickly and dramatically life would change and that life would never be the same again. Talk about breaking them down and rebuilding them! Matt came home and our whole world changed. It seemed every aspect of his life had suddenly turned green and we either got on board with the new rules or got yelled at – not seriously – at least the girls and I didn't take it too seriously. I'm sure Matt wanted us to be a little more on board than we were. I laugh now as I look back, because even his socks were folded in such a way that they had little smiley faces. Apparently, it's an Army thing. The entire Kapooka experience really did change him. He came home a different man to the one we knew before. He seemed stronger, more confident and, dare I say it, more grown-up. Gradually Matt became a soldier, and the more he got into it, the more he wanted.

Both our work commitments saw us away from home a lot. We became really good at playing tag team for the kids and scheduling overseas trips for me around training exercises for him. A training exercise is an Army term for going into the bush for a week or more at a time to practise war-like tactics. I'm not sure anything can ever really prepare you for being on the front line but this is more about giving you the skills to use the equipment correctly and sharpening your tactical skills. He would go away all clean and shiny, and come home smelling like crap and swearing like no man's business. It would always take a day or two for him to come down off the high of being away. He would come home and bark orders at us, moving a hundred miles an hour to ensure everything was in order and in its rightful place. Finally he would start to relax a little and settle back into civilian life.

Our so-called 'normal life' became an eclectic mix of Army life and civilian life. Trust me when I say the two are in complete contrast. But we just learnt to accept it, live the change and gradually it became our new normal. The kids loved the Army atmosphere. They thought going to the barracks was great fun and everyone was so friendly. I think they loved Uncle Matt in uniform too. It gave them a sense of pride and gave them a better understanding of what those today, yesterday and tomorrow sacrifice to be in uniform. I think we all got a better appreciation of what it means to be a member of the defence forces, what it means to be a part of that family and how to cope in the myriad of changes that life kept throwing at us. The kids went to school every day, they made friends, rode horses, did art classes and carried on with life as normal kids, and so it wasn't a huge shift for them at this stage. My work with CHEP was hectic but I thrived in it. I was away from home at least one week a month, sometimes two. Fortunately the girls were fine with Matt, and they both knew that if they wanted horses and pottery and music classes, the money to pay for those things had to come from somewhere. This was my livelihood and we all had to respect that, so we could all benefit from it. It was a great company to work for and my managers were incredibly supportive. If I needed a day off here and there for the girls, it was never a problem.

As for the Army Reserve, well, you see the adverts all the time: 'Join the Army Reserves, one weekend a month, two weeks a year'. My catchphrase for this quickly became: 'Join the Army Reserve, one weekend a month and two weeks a year is the only time you DON'T have to serve'. Part-time, my arse! Matt was away almost as much as I was, course after course, training exercise after training exercise. The Sergeant of the 12th/16th at the time was a full-timer turned Choc on a full-time service contract. This is known as CFTS (Contract Full-Time Service), which is how a Choc can become a full-time soldier without having to give the otherwise mandatory four-year service commitment. This meant his unit was run like a full-time unit even though it was actually a reservist unit. And so, it was nothing for Matt to be virtually full-time there as well. The Sergeant liked Matt and they worked well together. He knew how ambitious Matt was and he liked having people like him who were prepared to step up and be a soldier. Neither of us complained about the time and commitment. They paid him well, looked after him and, most importantly, we both knew he was a part of something so much bigger than us. He was exactly where he wanted to be and he was going to take every single opportunity that came his way, whether it was for a day or a month, he would say yes. He wanted to be the best of the best. Adapting quickly to change and displaying natural leadership qualities, it wasn't long before Matt caught the attention of his superiors, who worked with him to push him harder and drive him further. Anything Matt could put his hand up for he would do, and the unit loved him for it. He was determined to be as close to the perfect soldier as possible, committed, serious, dedicated and a great leader. He knew he had that potential and he also knew it put him on the right track to achieve his ultimate goal, to make it to SAS.

SAS – Special Air Services – is the pinnacle of soldiering, the elite, the best of the best. Only a small number of applicants make it through to the selection course each year and Matt was determined to be one of them. That was the dream and had been the goal since getting through basic training. He had finally found his feet and was wasting no time getting there. That's how Matt rolls; once he has locked onto something and wants it, he has to be the best of the best at it, not really competing against anyone other than himself. He puts a lot of pressure on himself and sets his own standards very high. This comes with highs and lows as he pushes himself to achieve whatever he sets his mind to. Being a reservist or even a full-time soldier was never going to be enough. He wanted more, and SAS was the only path he could see for himself that would fulfil that need. So, he locked on and set the path.

Within two years Matt went from Trooper Page to Lance Corporal Page. He was awarded Junior NCO (Non-commissioned Officer) of the Year and Soldier of the Year twice. He was promoted faster than most, probably because he wanted it more than most and worked his arse off to get there. He wanted to serve and he wanted to get overseas and fight. Everything about our world was changing rapidly and I knew at this stage that it was only a matter of time before he would go full time. It was like watching the bullet train, only instead of it passing me by really fast, I was a passenger and along for the ride. I'm not sure I associated injury or death or anything of a negative force with our family at this point; it's not something that is really at the forefront of discussion when a person becomes a soldier. I genuinely felt proud of his achievements; I felt happy that he was finding his feet and becoming the man he not only wanted to be, but could take pride in being as well. It suited him and it worked for us. The Army gave him a sense of purpose, something that really fulfilled him, and I loved being a part of that journey with him. It also meant that the Army slowly but surely seeped into my world. I was no longer just Kaz, with Matt and two daughters. Now I was Kaz, the soldier's wife who had to be respectful of the uniform and accept the seriousness of it all. I don't say this lightly when I say my husband was starting to bleed green.

Even today, after all we have been through, there is a piece of him that will always belong to the Green Machine. He was as committed to the Army as he was to me. I would never have asked him to choose, but if I did at that stage, I'm not sure which he would have chosen. He was all in and if he could have converted then and there, I know he would have in a heartbeat. By this time however, it wasn't me he was waiting on for the green light to go full time; he had to wait for the right opportunity. So for now, he would take what he could and bide his time. Although I can honestly say I was never the perfect Army wife, I loved him and knew I could never ask him to stop pursuing his dream. This was exactly where he wanted to be and it was beyond a want now, it had become a need.

Over the next few years we found a way to make our hectic lives work. We relied heavily on friends if travel plans for both of us clashed and, for the most part, we spent a lot of time passing in the doorway, either him leaving for the next training exercise or course or me heading overseas for my next project. We were essentially in a long-distance relationship and, when we were home, I was still working and he was still very committed to the Green Machine. He would be at the barracks any chance he got, sun-up to sunset. He took every job, no matter how menial the task, he was there, in uniform, doing the job. We also had the house and the kids to manage, but somehow, we just got on with it and made it all work.

We consider ourselves so lucky to have met so many wonderful people in Armidale who stepped up whenever we needed help. Looking back now, we would never have survived without the special network of people we met. Most of them are still in our lives today, on varying levels, but they all know who they are, and they know the special part they played in our story. Some were wives and partners of the Green Machine and others were civilians, but all of them loved us enough to help whenever we needed it. Due to the nature of our work, which meant we were sometimes called away at short notice, some of those friends would just drop everything and move into our home for a week to be with the kids. Others would take them to and from school, cook meals, care for the animals, the list goes on. Without the support of those amazing people, one of us would have had to give up a career. Thankfully we never had to make that choice.

The Army has a very strict hierarchy protocol and, not only do soldiers have to follow the protocol, but their partners are expected to follow suit as well, albeit to a lesser degree. As it turns out – surprise, surprise – I don't do rank well in civilian life. I am of the belief that if you want my respect you have to earn it, as I expect to have to earn yours. I'm no one special but I am also no one's doormat. If you bleed red and you breathe the same air as I do, then I don't really care what you do or how much you earn; you are my equal until you prove otherwise.

Meeting all levels of soldiers along the way, I learnt that some of them wake up every morning expecting that a rank automatically gives them the right to be a chauvinistic egotistical arrogant twat to anyone and everyone, civilian or not, and I won't respect that. However, there were some amazing soldiers along the way as well, more of them than the former and they, no matter their rank, worked to become my friend and they sacrifice every day in the name of the flag and uniform. That automatically has my respect.

For the most part I was pretty lucky. In my early introduction to Army life, the Sergeant in charge of the 12th/16th either loved you or hated you and your treatment was reflected as such. Matt, being as close to the perfect soldier as you would find, was loved. For me, that meant I was automatically accepted into the circle of the 12th/16th WAGS (wives and girlfriends). Some I'm still friends with today, some are no longer with their partner, as loving a man married to the Green Machine took its toll. Armidale WAGS were a lot of fun and I really enjoyed my time with them. I'm pretty sure most of the husbands cringed at the thought of a formal function because there was always one group of females up to no good, usually led by yours truly and a few others. The husbands would always hang their heads in embarrassment, but they knew it was always tongue in cheek. When we needed to abide by correct protocol, out of respect for the uniform, we always would. None of the Armidale WAGS ever brought rank into our world so we all got along really well and the social aspect in those early days was great. There was always a barbecue, a dinner or some other social event happening that brought the Green Machine family together. The more involved I got, the greater respect I had for the uniform and the people who wore it. It was never about the rank for me, it still isn't. Anyone prepared to put that uniform on and step up to fight a war and protect the innocent from the hell that war brings, has my respect. From desk jockey, transport officer and medico to front line gunslinging soldier, all deserve the respect of the nation. I only wish we, as a nation, we did a better job of showing it. It takes strength and courage beyond what most humans have, to be the person that runs towards the danger.

CHAPTER 5
It's just a contract

From 2004 to 2006 we found a way to make it all work and settled into a routine as best we could. The house was finally beginning to resemble more than a just converted tin shed and the kids were ticking off the usual growing up checkpoints and milestones. Jess was busy trying to find her passion outside of school. From music to pottery, drama to gymnastics, my eldest had tried them all. She had a flamboyant streak. In fact there was nothing shy about Jess; she loved performing and being the centre of attention. Ayleigh, on the other hand, was quiet and almost painfully shy. She didn't cope well in large groups and loved nothing better than being around animals. Spending most of her spare time on horseback, Ayleigh was happy as long as there was a horse in the backyard. Both girls had inherited my stubborn streak, and boy, could they fight! It was the usual sisterly antics, but they usually found a way to sort it out for themselves. I was still working from home for CHEP, travelling often, and Matt's time spent in uniform was increasing by the day.

Matt and I started talking about having a baby and we both agreed it was time. The more we discussed it, the more we both came to realise there was never going to be a 'right time'; there was never going to be enough money, enough stability, enough income, enough equity in the house, and the list goes on. Like everything else in life, we decided to just get on with it, set the challenge and make it happen. The idea that it might not happen for us never once crossed our minds.

We had been trying to conceive for about a year when Matt proposed. It's funny when I look back to that time now. He was already committed to me. There was never going to be anyone else for him and we were trying for a baby. So if that isn't commitment I don't know what is. He was neither here nor there about the marriage part, and after two failed attempts, you'd think I would have been gun-shy, but no, not me. I wanted the last 2%; I wanted the marriage. Maybe it's because I had failed so badly twice before, who knows? It's not like I had great role models to look up to; my parents spent more time fighting than getting along. But somehow I still believed in the concept of marriage. Maybe I just wanted a different ending with the right person. For whatever reason, I had my heart set on the happily ever after. Matt never understands it when I talk about the last 2%, mainly because he doesn't rate the whole marriage thing. As far as he was concerned – and still is – marriage is just a certificate, a contract. Mind you, he wouldn't undo any of what we did. And so, knowing how much it meant to me, he settled for the 'happy wife, happy life' option and popped the question in February 2006. Together we bought my first ever engagement ring. Yes, despite two previous marriages, it was my first: a beautiful champagne diamond, bezel set with three smaller stones. Being a bit of a traditionalist at heart, Matt also decided to ask for my dad's permission. Like my dad was ever going to say no! All jokes aside though, I was secretly chuffed that he even gave my dad a thought. Dad had come to terms with our relationship a while back and once he took the time to get to know Matt, he discovered what most do, that despite his cocky exterior, Matt is a genuinely good guy with a heart of gold.

And so we packed the car and headed to Sydney for a long weekend at Mum and Dad's. I was really hoping Matt would just get it over with quickly. Dad had this routine where he would march off to the club every afternoon at five to meet the boys, and I wanted Matt to get in before he left. On our first of three nights I waited patiently for Matt to approach Dad. I watched the clock in the kitchen and waited. Five o'clock arrived and off Dad went to the pub. Nothing. Damn! Day two and I waited again, watching the clock with less patience than the day before. Yes, it has been mentioned to me once or twice that I can be a little controlling in situations, especially if the situation directly affects me, and no, in forty-something years I have not managed to polish this particular aspect of my personality. So, on this particular occasion, I insisted that my dad hang around after five. As the minutes went on, Dad became increasingly fidgety and frustrated. When he pushed to know why I needed him to stay, I told him it was because Matt wanted to talk to him. As it turns out I did myself no service with this tactic and it all went horribly wrong. When I told Matt that my dad was waiting for him, it went down like a lead balloon, and rightly so. But he went anyway.

Dad was having a beer when Matt casually said, 'I'd like to have your permission to marry Kaz'. Dad looked at Matt in disbelief and then burst out laughing, nearly choking on his beer as he spluttered, 'What the bloody hell for?' Regaining his composure, he then said, "Yeah sure, but I think you're a bloody fool and there's no refund'. I heard all of this unfold from the bedroom down the hall. To this day I'm not sure how much of that was tongue in cheek and how much was serious. With Dad's 'blessing' Matt got up from the table, walked down the hallway, saw me sitting on the bed and nonchalantly threw the little velvet box at me, 'There you go, your Dad said yes! You can go ahead and wear it now'.

We told the world we were engaged and started making plans for a wedding, but inwardly I was pissed off and the more time passed, the worse it got. I was angry at how it all turned out even though I knew it was my own fault. But at the time that didn't matter; I just wanted it to be different. It's like when you build the memory before the actual event takes place and then the memory is shattered because the event doesn't unfold the way you want it to.

Suddenly I had an epiphany. Matt could buy me another ring and ask me to marry him properly! To Matt's credit, when I approached him with my idea he didn't brush it off or remind me of the fact that I had ruined the original moment myself. To Matt's credit, he went along with it, found a new ring, a white diamond this time, and replaced the old memory with a good one. I won't go into the details, but let's just say the new memory still makes me smile, blush and giggle a little, even ten plus years on.

MATT'S PERSPECTIVE

The thing that makes Kaz so formidable in life is her determination and willingness to prevail, no matter what gets in her way. So when we arrived at Brian and Net's place, Kaz was champing at the bit for me to ask her father in her time, not mine. When she managed to talk Brian into delaying going to the pub I wasn't thrilled about it and responded in a mature and responsible manner: by ignoring her and storming around the house! Really mature, I know, and I'd like to think over time I have outgrown of this kind of behaviour, especially because no matter how much people may tread on your toes it's not a great way to treat the people you love.

Kaz's recount about what happened is pretty accurate. I'm not defending the way I acted; I was just really hurt. I'm pretty old school and it was super important to me to do this right, especially considering Kaz's history and the fact that I hadn't been married before. So when things didn't go the way I expected they would my pride was hurt. I did end up asking Kaz to marry me again and had to purchase a new ring to boot! Ultimately, I am a big softie at heart and if that's what it was going to take to make her feel right about the situation I was going to do it.

Planning the wedding was a challenge in itself. For starters, where to have it? A church? We met with a priest, but as a two-time divorcee, my God, they needed a truckload of paperwork to let me marry again! And in all honesty, even though both Matt and I were christened Catholic, we knew they were never going to be happy to marry us. And we weren't, aren't and never will be practising Catholics anyway. So we opted for a civil ceremony. Now the question was where to hold the ceremony. Where could we bring all our friends and family together to celebrate our marriage? We were talking about it one night when Matt turned to me and suggested, 'What about having the reception at the Army barracks?' It made sense given how entrenched the Green

Machine was in our lives by this time. Matt was now working full time with the 12th/16th, despite still being a reservist. As the boss' 'golden boy', it came as no surprise to me when his request was immediately granted. The Sergeant even offered to put a bottle of red and a bottle of white from his private collection onto each table as a wedding gift. As for the wedding itself, we decided to hold it in the local public park with a celebrant to take us through the ceremony.

With both of us still working full time and travelling often, it took about ten months to bring it all together. We were busy but overall pretty settled. Matt was focused on his long-term goal and even though we didn't speak of it often or deeply, we both knew where he was heading. It was like the unspoken future. We were also still trying for a baby and, as each month rolled into the next, without success, the thought that I was failing to produce a child for the man I loved, the man I promised a child to, began to niggle at the back of my mind. Fortunately I was mostly able to bury those thoughts behind organising the wedding, working and being a mum to Jess and Ayleigh. The girls were excited about the wedding too, and genuinely happy to see their mum and Uncle Matt tie the knot. I guess it gave them a greater sense of security too. We went wedding dress shopping together and they helped me plan the music, gave me feedback on the invitations and helped make the table decorations.

Our wedding was made possible with the help and generosity of family and friends. My cousin was our photographer, his wife made the wedding cupcakes and the invitations were designed and printed by a friend. I selected Jussy (no surprise there), my two girls and Mellissa, a friend I met in Armidale and instantly clicked with, as my bridesmaids. Melli is one of those people who gets along with just about everyone and calls it as she sees it. She is hilarious to be around and when my dad would come for visits, Melli and Dad would play card games and drink beer at 10 o'clock in the morning. They shared the same dry sense of humour and got along like a house on fire. I remember Dad asking her where she worked on one of those occasions and she responded with, 'I'm an LJ Hooker's girl'. Dad thought this was hilarious and referred to her from then on as 'Hooker', all tongue in cheek, of course. 'How's Hooker going?', he would ask in reference to Melli. It just stuck.

Matt chose Ronnie, Jason, who was an old schoolmate, and Dan, the Army pastor, who he had become close with, to stand by his side. Our guest list was a combination of CHEP, Army, family and friends. They were an amazing bunch of people who had been around for a long time, a short time and everything in between. There were people on the guest list that I hadn't even met. Holding our reception at the Army barracks meant we were pretty much inviting the entire barracks and their partners. I was yet to meet all the partners but figured our wedding was as good a place as any.

After what felt like an eternity, the big day finally arrived: the 1st of September 2007. From the moment I opened my eyes I knew it was going to be an amazing day. Getting ready, I decided to take it slow, savour every moment. I didn't want it to be a blur down the track and not be able to remember the details. I'm so glad I did that now, because I can still remember that morning so clearly, how happy and excited I was and how genuinely happy our family and friends seemed to be for us. I wanted this one to be the last and I wanted it to last.

It was the best wedding I could have ever hoped for. I drink socially and most of the Army folk and our friends and family are also known to enjoy a few, so the celebrations continued on into the evening and by the end of the night there were just a few hard partyers left with Matt and I. Matt doesn't drink, he stopped not long after we met. He just doesn't find any interest in alcohol. But he is usually so full of life that you can't tell he is the only sober one in the room. When Matt and a group of his drinking mates get together, it usually ends up in shenanigans of some description. For those of you who have ever been in, had a family member in, or just know someone in the Army, you would know that those boys and girls know how to party. And boy did we party!

If I were to call out one slight glitch in the entire event it would have to be the morning after our amazing wedding. The next day was Father's Day, so I had organised for all the guests to join us from 7:30 am at our tin shed for a Father's Day breakfast and a toast. Our first morning as husband and wife, of wedded bliss. Waking up feeling happy and snuggly and so in love, I rolled over to face Matt and said dreamily, 'Good morning, husband', to which he responded, 'Fuck off!' Needless to say it was not the response I was expecting. Shocked and taken aback, I questioned him on his abrupt and rather nasty response. It turns out that my totally sober husband was not too happy when he discovered that our house was in a complete shambles after the party was over, especially after bringing me, his very drunk and disorderly wife, home only to have me waltz through the house, undressing as I went, before climbing into bed and falling asleep the second my head hit the pillow. This would normally be fine if it weren't for the fact that I had invited about forty guests for breakfast early the next morning. Apparently five women, all getting ready for a wedding at the same time and in the same place, can make quite a mess, which my adorable husband cleaned up while I slept. It was the early hours of the morning when he finally crawled into bed next to me, so it must have felt like only minutes had passed when I woke him with my cheery good morning. I still laugh about it now. The house was immaculate, and brekky was a huge success. Thank you, husband!

MATT'S PERSPECTIVE

The lead-up to our wedding was pretty much like any other, as far as I could tell anyway. The bride-to-be organises 90% of it, and probably for good reason. What would any of us men know about doing anything remotely related to a wedding let alone get it marginally right in the eyes of our soon-to-be wife? So in order to make it to the day and beyond, I did as any other groom would do and gave my bride free reign to sample, interview, phone, email, and – within reason – shop until she was satisfied. So my job was pretty easy: try to show interest, even if I couldn't tell the difference between the invitations in Antique White or Hog's Bristle (both looked white to me).

Here's my advice to other grooms: don't ask what the big deal is, look as though you're interested and the choice is important to you and then give your bride-to-be the final say. Be enthusiastic and turn up to appointments. Otherwise, just stick with organising the suits and a wedding ring and writing your vows. Then twenty-four hours out make yourself scarce and leave the women to do their thing.

I can't speak for other blokes, but you hear the usual things about pre-wedding nerves and cold feet but, to be honest, I felt none of it. I knew where I had to be, what I had to do, and when. And sure enough it all went off without a hitch. Another one of Kaz's strengths is her ability to pull things together; she has a real mastery of this skill. Wedding done, on to the photos, done, then on to the reception. When we arrived at the venue everybody was there waiting, and it was pretty obvious that they'd already been partaking of the bar. We had an amazing night surrounded by friends and family and everything was pretty well wrapped up by 10 that night.

As we said our goodbyes and made our way to the car, I became aware of my now wife's inebriated state. So I assisted her to the car and drove us home. I pulled up to the house and sent Kaz inside so I could gather up all the gifts. As I struggled through the door my wife had absently shut on me, my arms full, I nearly tripped over the clothing she had shed on her way to our bedroom. I unloaded the gifts, gathered the clothes and walked towards the bedroom. Now, normally this would be the part about the romantic evening the newlyweds spent on their first night as man and wife. Well, not this story! I walked in to see my wife had thrown herself face down and bare-arsed on the bed amongst what appeared to be half the towels in the house, snoring like a Kodiak bear with a head cold.

Gathering up the towels and taking them to the laundry, I decided to venture over to the other side of the house where I found the remaining towels, what appeared to be the contents of three makeup kits, two hair dryers and a cornucopia of tools, devices, clips, pins, brushes, sponges, as well as at least three empty bottles of bubbly with accompanying glasses. I took a deep breath, rolled up my sleeves and got stuck into the cleanup. By about 1:30 that morning the house was in a reasonable state for the forty or so guests we had turning up for breakfast. I crawled into bed and closed my eyes. After what felt like only five minutes, I was abruptly woken by my wife with a boisterous and overly cheerful, 'Good morning, husband'. After giving her my account of what had happened during the night, she simply laughed and said, "Oh babe, I love you', repeating those words over and over until I reluctantly said it back to her so I could close my eyes for a bit longer. Forty-five minutes later, the first guests arrived.

Misdemeanours aside, one thing I can unequivocally say is that Kaz pulled it off. We had the wedding and reception of a lifetime. It was typically us, surrounded by people we love in a venue that probably had never been used for this purpose before and probably never would be again. We had an abundance of good, simple food, everybody was able to drink their share, and we made memories that would last a lifetime. And to top it all off, Kaz brought it all home for under $10,000, which was no small feat for a wedding that had over 120 guests.

While we were working our arses off, planning our wedding and trying to fall pregnant throughout 2007, something was simmering in the background. What had begun as whispers and rumours that the Army would offer Contract Full-Time Service (CFTS) for Chocs to join Brisbane in a deployment to Iraq suddenly became very real. I knew there was no way Matt was going to miss out on a deployment. He had asked me ages ago whether or not I would support him if he ever got the chance to go on a tour of duty and I had said yes, not really knowing at the time what that would mean. I can still see the look in his eyes when he asked me, a look that told me how much it meant to him. As there was no guarantee that he would get the contract, we didn't talk about it much. But it was there, always just off centre, like a dark cloud hanging over us. For some reason, even though I knew deep down that it would happen, I chose to push it out of my mind. Our wedding was way more important, and having a baby would change everything anyway.

CHAPTER 6

My husband went to war

With the wedding and Father's Day over, everyone went back home and we settled in to life as a happily married couple. The only disappointment was that we didn't get to go on a honeymoon. I still joke that our honeymoon was Matt going to Iraq.

It was no secret leading up to the wedding that Matt was putting his hat in the ring for a deployment. The boys would largely refer to it as 'a job', and Matt had to undergo some intense training to ensure he got it. Some had their doubts about him landing an overseas 'job', but not Matt. Not once did he doubt it. He wanted it and he was going to do whatever it took to get on that list. We talked about it and I was excited for him. But in all honesty, I hadn't properly thought it through, including the impact it would have on the girls and me. It was what he wanted and he was so passionate about it, how could I not support it? It's what they train for; it's ultimately what it's all about. Go to war and protect those that can't protect themselves. That sentence right there could divide a nation, I'm sure. I know firsthand what a lot of people think about our soldiers going to war overseas. I have heard many forced opinions over the years. One of the joys of being an Army wife is that people are so quick to tell you why the boys should or shouldn't be there. My response to the haters has always been the same and very simple, 'The boys (and girls) didn't pick or cause the war, the men in suits that you voted for did. If you want to be angry, go be angry at them, and remember this, if war lands on our shores, it's those men and women in uniform you will be hoping will save our country, save your freedom and ultimately save you. So respect our soldiers, don't blame them'. This response would usually and swiftly bring silence to the conversation.

As strange as it may sound, Matt wanted to go to war. I don't think they really process it as war, not really. I think these guys just want to put everything they train so hard for to use. Matt got his wish. A week after our wedding my brand new husband was signed to a twelve-month Contract Full-Time Service (CFTS) to Brisbane and walking out the door to leave for pre-deployment training. That week leading up to the day he left felt like an eternity. Things were very subdued between us. Suddenly, everything we could hide behind to ignore the reality of him going to war was gone. He was leaving, and I had just seven days left with him. We would get to talk over the phone occasionally during his pre-deployment training and we would have a few days together right at the end just before he deployed, but right then in that moment, the time when most are in honeymoon mode, still celebrating the thrill and excitement of being married, our house was very quiet. The realisation that he was going to be away for so long was heavy enough but on top of that, I had not yet fallen pregnant. It was going to mean another eight-month delay, eight months of not being able to try. None of the doctors we saw could tell us why it wasn't happening for us. The wedding had given me something to bury myself in, a distraction from my inability to fall pregnant and a distraction from the idea of Matt going to war. But that was all over now, and I had nothing but time to think about. I tried so hard not to cry during that week, not to make it harder for Matt to leave. I tried not to let the girls see how sad I was, crashing so hard and fast on the back of the incredible high of our wedding. Matt was great, he was supportive and lovely but at the end of the day, the reality of it all had kicked in and I was sad, I was scared, and I had no idea how to process any of it. It was hard for the girls too. They had to watch it all unfold, not really understanding the impact of it all, although they did seem to get that he was going away for a long time. I felt bad for them having to see so much sadness, but kids are far more resilient than we usually give them credit for. In the moments I couldn't hide it, they just hugged me and told me it was all going to be fine and that Uncle Matt would be home before we knew it.

Matt spent a lot of time getting ready, double and triple checking that he had packed everything he needed. He would pack, unpack and repack multiple times. I think this was as much about mental preparation as it was physical packing. He is a very quiet person most of the time, a deep thinker, always observing, listening, learning and absorbing information. This was new to all of us. He had to figure it out in his own way and in his own time, while also being there for two stepdaughters and a wife who had no clue what this was all going to mean. The Army doesn't really give you insight into this. I think they could do so much better for the families when soldiers are getting ready for deployment. The 12th/16th was great though. The same Sergeant was still in charge and he was awesome; he made sure Matt knew that we would be looked after and that if I needed anything, anything at all, all I had to do was ask. And I had some amazing WAGS around me at that point and some great civilian friends too. I wasn't going to be alone and Matt didn't need to worry about the girls and me on our little farm. He could leave knowing we had people there to support us and help when and if we needed it. So by the end of our first week of marriage, Matt had his paperwork, his bags were packed and he was walking out the door to learn how to be a front line soldier.

Matt spent the rest of September, October and most of November away, preparing for Iraq. All up he was to be gone eight months, two months of training and six months on the front line. He came back home for four days just before he was to fly out of the country. During those precious days I distinctly remember one particular incident that caused a reaction from Matt that was out of character. He was packing when he suddenly realised he wasn't wearing his wedding ring. Everything was so neat and tidy and in its rightful place, but no wedding ring. He looked frantically, trying to keep everything in its place, but no luck. So, next minute there was Matt pulling everything out of his bag. He went from organised to chaos in about five minutes flat. He was so worried he had lost it, this thing that symbolised something he was neither here nor there about before. He eventually found it. Evidently it had fallen from our dresser into one of the drawers. Visibly relieved, off he went to repack. After he had calmed down and finished repacking, I asked him why he had been so worried about the ring. He said he didn't want to go away for so long and not have the ring on. He knew how much it meant to me and he knew the seriousness of what he was asking of me to support his decision to go, so he didn't want that to be any harder on me. He told me he had taken so long to find the right wedding ring for himself because he was only ever going to have one, and that what the ring symbolised for me, it meant for him as well. Right answer, husband – well played!

MATT'S PERSPECTIVE

So how's this for a honeymoon! One week after we married and I was on a plane to Puckapunyal in Victoria for a vehicle course with the Army after receiving word that I had been offered a place on a deployment to Iraq, departing in November. In addition to the vehicle course, I had pre-deployment training to complete before I left so I wasn't home a lot, which meant it all fell on Kaz to manage the kids and the house. I must admit back then I was so driven and focused on my goals, I never really acknowledged the weight and sacrifice my loved ones were making so that I could achieve my professional goals. True to her nature, Kaz was totally onboard. She moved heaven and Earth in order for all of this to happen. At the same time, we were doing IVF to conceive, which was a massive undertaking and not one I would fully understand the ramifications of for Kaz till much later. When I left for deployment, it was all weighing heavily on her, not that she let any of it show.

I left for Iraq and it was an exciting time, because, as far as I knew, I was heading off to conduct my role in a live theatre. Little did I know that this would not prove to be the pinnacle of my career, as I had hoped it would be. I was deploying with a team of mostly near-geriatric officers and warrant officers, who had been around the Army so long they had to be pulled out of the archives in order to deploy. We were there to help train Iraqi soldiers at the Recruit Training Centre. The reality of the situation quickly revealed itself; it was everything soldiers today complain about. The senior officers and warrant officers I was deployed with had their own ideas of how things should be. Each thought his idea was the best and that he was the only one capable to deploy. They were constantly throwing their two cents in on anything my team and I were trying to do, despite the fact that we were all proficient senior soldiers in our own right with more knowledge in our individual fields than they had. We also provided their lift capabilities and security. To top it all off, everything we did, all the training and leadership and effort we put into the Iraqi soldiers, ended up benefitting a little group called ISIS (I'm sure you've heard of them) where a good number of them ended up defecting. Was it all worth it? In hindsight, no it wasn't; not when my wife was going through hell with IVF while still having to travel for work and look after the girls and the property.

It was at this point that I started to see through the veil. I began to discover that, while the Army is an organisation consisting of some of the best people in the country who possess the highest tier of capabilities, it is also rife with self-serving loathsome people who are all about their career and not at all about good, honest, capable soldiering. I started to see the toll that these individuals would have on others. I saw top-level professionals worn down and fractured mentally because of the absolute self-serving, self-aggrandising nature in which some members conducted themselves. What I would later come to realise is that the mental health of our soldiers is greatly affected by those around them, especially those in command positions. A soldier is seriously affected when he is 'cast out' by his peers, has backs turned on him or is consistently doing more than his fair share without acknowledgment or respect. This is also the case when soldiers' lives are put at risk due to pettiness of the commanding classes. While the commanders are there to be obeyed, they also have to take responsibility for their actions and have a duty of care for the men and women who serve them, and who they serve in return.

For all this I left my wife, step-kids and potentially a future child. It has taken me many years to try and get a hold of this part of my nature. Hopefully I am getting closer to mastering it and really valuing my family, especially Kaz, who has been extremely patient, and hopefully has enough in her till I manage to get it right.

On the 25th of November 2007, Lance Corporal Matthew Page left for a six-month deployment to Iraq and no one could have prepared me for what lay ahead. The tears I shed were for so many reasons but mostly because my husband was heading to a war zone, the fear of what may happen, the idea of being separated for so long, the lack of understanding and not knowing when we would speak, for how long or how often. The day he left was the day I wanted him back home. I wanted to scream at him, 'No, no you can't go and be a soldier in a war zone! No, you can't go to another country where I can't be with you for six months and who knows what will happen to you'. But I didn't. I told him I loved him and that I would be right here waiting for him until the day he comes home. I asked him to stay safe and to remember he had people back home who needed him. And I promised him that I would record what I was doing at home, because all being well, he would come home a father.

CHAPTER 7
Battles of a different kind

For two years leading up to this moment we had tried everything possible to fall pregnant. We had been poked and prodded and tested. We had been given hormone drugs and fertility drugs and even placebos. Nothing. And we weren't given any reasons why it wasn't happening for us, no medical reasons anyway. It brought constant doubt into my head. Would we ever have children together? All I could think of at the time was, 'Oh my God, if I don't fall pregnant, what have I done to this guy? How will I live with the fact that I will never make him a father?' Watching his love and devotion to my girls, fully aware that they were never really his and that another man would always be called Dad by them, was heartbreaking. To them he was Uncle Matt, and he was amazing, and he loved them. But he wasn't Dad. How could I ask him to spend the rest of his life with me knowing that I may not be able to give him the one thing he really deserved to experience and really wanted? And he wanted so little. It fucked with my head. Matt was amazing; he just believed it would happen and everything would be fine. He used to tell me if I would just relax and be patient, it will happen. That was like asking me to stop breathing! I knew if I took my eye off the ball and relaxed, we were never going to fall pregnant. Turns out, occasionally I'm wrong; just don't tell anyone... especially Matt.

MATT'S PERSPECTIVE

I knew it would happen when it was meant to happen. Even though she forgets it at times, especially when she gets scared, Kaz knows I have a good natural intuition, and my gut was telling me, 'Relax, you guys will have a baby when the time is right'. But Kaz was never convinced, no matter how much I tried to explain that I could just feel it was going to be okay. But when your wife is in tears and hating life as it stands, trying to tell her it will be okay goes down like a lead balloon and I just couldn't add to the rampant emotions floating around inside her head. So I chose to just support her through this process as best I could, even when that process was happening while we were both in different countries.

Just before Matt left for Iraq we froze his swimmers. He still hates me referring to it like that but that is what we did.

MATT'S PERSPECTIVE

Piece of advice: if you ever find yourself in this situation and go into one of those little rooms, be careful when you turn on the TV. I turned on the TV in my little cubicle because, when in Rome, right? And what do I get but an image that, let's just say, didn't help the mood! The previous occupant of the room either had very different tastes or was trying to mess with the next guy.

We had been discussing having a baby for about six months leading up to his departure. We both hoped it would just happen naturally and that we wouldn't need to go down the IVF path, but it didn't. Matt was sure it was more to do with lifestyle, with both of us working so much and being separated so often, but I wasn't convinced. With the separation looming, I knew it was only going to get worse. So I wanted to look for alternatives. One of the doctors we were working with had discussed the option of IVF a while ago. We kept putting it off. But it was time. So after further discussions with the doctor, we put the wheels in motion. We had a plan that meant we were going to have a baby via medical intervention.

So Matt went to war and I went to the IVF clinic. I went through the entire process on my own, while my husband was in Iraq. I had fantastic friends around me, and I was thankful for that. However, anyone that has ever been through IVF will tell you that an important part of the process is having your partner walking through the journey with you. There are so many highs and lows with the process. For weeks and weeks I had to stab myself and inject hormones and deal with the side effects. They touch on the side effects, almost as an afterthought. Your hormones go wild while life continues, and somewhere in amongst it all you have to try and find a balance. It was like being back at the University of New England all over again. Crying for no reason – only now I had about ten reasons – happy one second, world crashing around my feet the next. I tried my best to hide the rollercoaster of emotions from everyone and to keep everything as normal as possible. I would send Matt little video recordings of me injecting myself, just to somehow have him be a part of it all in some small way. I tried even harder to ignore where he was and what he was doing, as I desperately wanted my baby to have a dad. I kept telling myself everything was normal and it would all be okay. But in reality I was no longer certain what 'normal' was in my world anymore, or even if there had ever really been a time when things were 'normal'. Here I was, a newlywed two-time divorcee. I had just married a man who made me laugh like no one else ever has and we were supposed to be going through this amazing baby-making journey together. Instead he was in a war zone and I was going through IVF without him. I liken how I felt to holding your breath every day. The pressure builds and builds until you must let go and breathe. That's what it was like, waiting each day for some sort of communication from him. Sometimes it would be days or even a week before he could contact me. Try holding your breath for a week – and throw in a few hormone injections while you're there!

I'm not sure how much Jess and Ayleigh understood about what was happening and they certainly had no idea about the impact it was having on my hormones and feelings. They knew their step-dad was in a war zone and that was enough for them to have to deal with without me dragging them through the turmoil of what I was going through as well. So I never told them, not entirely, and I'm not sure if that was the right thing to do. I thought I was protecting them. I thought crying every night after they went to sleep was better than crying in front of them. I just wanted to be strong for them, I wanted them to feel like everything was going to be okay and that Uncle Matt would be fine, and we would have a baby and all would be right in the world. Yes, I know how stupid that all sounds now, but in the thick of it, I just went with what I thought was best for them, which was to tell them as little as possible so they wouldn't have to worry, put on a brave face every day and have my private emotional breakdown every night. Sometimes I didn't make it to the night. Sometimes I got as far as seeing the girls off on the school bus and then I would come back inside and cry for a bit, take my hormone injections, cry a bit more and go to work.

When you speak to 'the boys', they all play it down: the impact, the reality of the danger they were in on the front line. But it is real, soldiers did die and each one of them had a family, a wife, loved ones, kids, all the people that needed that soldier to come home. And now what's left for that wife, those kids and that family? A memory of a man who was everything to them but is now just a memory, a shattered world in which they need to put the pieces back together and make a new path, all the while trying not to let the words, 'should have' break them. This man that should have come home to them, becomes a brief news article, a passing comment from someone they used to know, an afterthought to most, a face on ANZAC Day and Remembrance Day. Lest We Forget. They were, are and will always be so much more.

That harsh reality had become my world. Every day that ended with no official car in my driveway was a good day. It's an intensive way to live, and in some ways, you just have to bury it and move on. I still had my work to focus on during the day, and CHEP was amazing. My boss at the time was so supportive and the company knew me, they knew my work ethic and knew I would do whatever it took to get the work done, even if it wasn't always within a structured and mechanical nine to five existence. For all intents and purposes, I was also a single mum. Each day I just got on with living and each night I prayed that the demons of fear would be a little kinder and let me get some sleep. However, most nights they were anything but kind; the tiniest noise would make me stir and I'd be wide awake. Then in the depths of silence the fear would creep in and reality would sink in. He may not come home. This may not work. How the hell do I get through this?

IVF is an interesting process. It fills you with such hope and yet all the medical professionals tell you not to get too excited and that sometimes it doesn't work. But they only say that once and follow through with a big smile saying, 'But you'll be fine', so you cling to that optimism. The process itself sucks, quite frankly. You have to inject a round of hormones into your stomach daily for a period of time and then go to a clinic in Sydney to have as many over-stimulated eggs as they can access removed. On the first of those occasions, my friend, Sarah, offered to come for the ride with me. It's a four-hour drive down to Sydney from Armidale, probably closer to five hours into the CBD. We arrived a little early, so we had some lunch at The Rocks. It was a lovely day and I was really excited at the wonder of it all. Once the eggs are removed, they mix in a bit of Matt to make an embryo, and hey presto, we have a baby! It really is a miraculous process.

One thing I discovered is that they downplay is the pain. They told me I would feel some discomfort, but holy shit they definitely underestimated that one! I had already had children so I knew what contractions felt like; this was definitely up there. And it was no better after the procedure; in fact, I would say it was worse. I was in agony and had to endure the five-hour drive back to Armidale. Thank God I had a friend with me because there was no way I could have driven myself home. Despite the pain, it ended up being a good day because they were able to extract quite a few eggs. So I sucked it up, put on my big girl pants and fell asleep in the car while Sarah drove us home.

Soon afterwards I received a call that all had gone well and we had three viable embryos. I was ecstatic! I made an appointment for embryo implantation a couple of weeks later. I couldn't wait to hear from Matt so I could tell him the great news. The communication with him was sporadic and even when he could call, it wasn't a great line and he never had a lot of time to chat. There were so many of them over there and only a few lines to call from, so time was limited in respect for others waiting. Eventually I got the call and was able to share the news. Matt was excited too, although I believe more for me than anything else. He figured we were having a baby one way or another, so he wasn't worried. But he knew how much this meant for me.

As I had 'spares' the doctors suggested I go with a single implant first. This time Valerie accompanied me on the long drive down to Sydney. This visit was a completely different experience to the last. Here we were, my girlfriend and I in a clinical white room with a doctor and a nurse. They had me reclined on a lounge type hospital bed and there was a massive screen on the wall so I could see our baby (embryo) being implanted. In no time at all the embryo went from Petri dish to my womb, and bang, I was pregnant!

So here I was, pregnant while my husband was at war. That's a journey filled with all kinds of screwed-up emotions, let me tell you! Some days I found it hard to tell if my tears were of joy or sorrow. Many times I think they were both. I wasn't prepared to be a war wife and an IVF statistic at the same time. No one had really helped me understand the impact of Matt going to war. When he had been assigned to a unit in Brisbane, I was still in Armidale with a unit of peacetime soldiers. No one really knew what to do. They were all amazing at the 12th/16th and rallied around me as best they could but there were no WAGS that really got it because none had experienced it. None of my civilian friends had a clue either. Please don't get me wrong here, I feel so fortunate for the friends I had and without them this story would have been very different. But there were times when I just wanted someone to get it.

Once an embryo is implanted, there is a delay of about a week before you have blood tests and an ultrasound to confirm the pregnancy and ensure everything is as it should be. That week felt like one of the longest in my life. My emotions were all over the place, a real mix of excitement and nerves. Every now and then I allowed myself to believe that this could actually be real, and I could actually be pregnant with Matt's baby. I remember going to sleep at night holding my stomach, willing with everything I had for the baby to be okay. 'Grow baby, just grow; it will be okay. We will be amazing parents for you, and you have big sisters that will be there for you as well'. Every night, same chant, just willing this baby to grow.

'No sign of life'. Those were the words I heard as the very kind nurse on the other end of the line reported my blood test results, which indicated I had lost the embryo. In fact it had never actually taken. I wasn't even pregnant! That's it, just a phone call letting you know that the blood test showed no sign of life. The end. I was all on my own and I was gutted. I thanked the nurse who very kindly reminded me that I had more embryos so I could try again. Like fuck I'm trying again! My heart had just been ripped out, jumped on a thousand times and thrown back inside.

MATT'S PERSPECTIVE

I remember getting an email from Kaz's friend explaining that I needed to get in contact with Kaz and that I should really be a little more thoughtful about her situation, especially considering the IVF failure. To say it angered me would be an understatement. In my mind I responded to that friend with sarcasm, 'You do get where I am, don't you? Do you think I would have my potentially pregnant wife go through this and not call her?' I had literally just returned from a job, gone in to see if my wife had emailed me and this is what I get! In hindsight, Val was just looking out for Kaz and didn't really have any idea about how operations in the Middle East work. So I pushed down my thundering anger and phoned my wife who needed me.

Many, many more tears were shed and none of them were for joy. I was broken, Matt wasn't here, and I just didn't know how to move forward. It's like your feet are frozen and life is just too heavy. You can't run, you can't move, you just have to stand there, cop the shit thrown at you and deal with it as best you can. I had two beautiful girls dependent on me, so I had no choice but to put one foot in front of the other every day and die a little inside every night. I was so sorry. I was sorry for myself, I was sorry for our loss, but most of all I was sorry I failed Matt. I didn't take the nurse's advice to try again. Those embryos stayed put in the Petri dish for the rest of Matt's deployment. The pressure to give him a baby and get through his deployment at the same time was too much and I still had another three months to go. So the girls and I just got on with life, hoping every day that Matt was coming home alive and in one piece. I'd lost our chance to have a baby. I was riddled with guilt and broken as it was. I wouldn't have had the strength to move forward if I lost him as well.

MATT'S PERSPECTIVE

It was about this time that there was an incident that resides with me to this day. At the entry to the air base where we lived was a big pair of arches. On this particular day, a contractor's car was approaching the arches, preparing to enter the base to commence the day's work. The contractor was coming through at roughly the same time we were due to leave. Suddenly there was an IDF (indirect fire) attack. This is when local insurgents use basic means, like structural steel or a tool similar to a car jack, to launch stolen or illegally sought rockets towards their unsuspecting victims. This particular incident was unique as far as attacks go. More than thirteen rockets hit that day, which is very unusual. It occurred during the morning, also unusual. It hit a moving vehicle, which was quite deep within the outer perimeter – again unusual – and hit them as they were heading to work, just kicking off their day. Something like this makes you realise just how random and unforgiving death can be. It will grab you against all odds and it will grab you whether you're sleeping soundly or smiling with your mates on your way to work. It made me realise you cannot live half a life. You have to grasp life and wring every pleasure and joy you possibly can from it. I never told Kaz about this incident. In fact I kept most things like this from her to protect her, unless I thought she would hear it on the news. In fact, I do believe reading this recount will be the first she learns of it!

People often said to me, 'Wow, that must have been so hard. How did you not stress out every day?' I'm not sure what made them think I didn't. Of course I did! I knew where he was; I knew what might happen. I'm not stupid. It was present in every thought I had. Even though I didn't want it there, it was like a white elephant in the room. Every time I met up with someone – friends, family, or colleagues – they all knew exactly where my husband was and what he did and they knew I might be struggling. And because they cared about me, the first thing they would always ask was, 'How are you going, Kaz? Have you heard from Matt? Can we do anything to help?' Of course I couldn't be mad at these people. I knew they were asking because they loved me and were genuinely concerned for me and the girls, and Matt. Most of them would have done anything to help make it a little easier for me. A lot of them did. But sometimes, just very occasionally, I couldn't help but wish they didn't ask. I hoped they would just see that I didn't have the strength to answer. I didn't want him to be there. I wanted him home with me, not over there, fighting a war, risking his life, risking our lives together for a war neither asked for nor wanted. Some days, I was so mad at our country for sending him to war and even madder at him for choosing to be a soldier. Some days, I just wanted to be entirely selfish, have it all be about me and not think about him or where he was or what he was doing. Turns out, it's much easier to just shut down and answer the questions.

Christmas proved to be a nice distraction as we went to Sydney for a family reunion of sorts, but I needed more than just a short-term distraction to get through the next couple of months. I decided I needed something to dive into and focus on, something to take me away from the constant negative chatter in my head. Matt was going to come home and I wanted the world to know, well at least the Armidale/Tamworth world, that he and the boys were heroes. So, I got busy making as many contacts as I could to ensure the media, officials and everyone we knew would be there to welcome our heroes, including my hero, home.

CHAPTER 8
The soldier comes home

Matt finally returned home in May 2008 to a fanfare of family, friends and media at Armidale Airport. I had, at that point in my life, never been so relieved to see someone. He was home and he was safe. Nothing else mattered. All the other shit that comes with a returning soldier could be dealt with later. He was home and he was safe, and I was in his arms, which meant I was safe too. It's interesting how the Army trivialises all that other shit. They actually gave me a handbook. It read something like this: 'Hey wife, when your husband gets home, he may struggle a little to assimilate back into normal life with you and the family. He may not be the same person he was when he left. He may need space, etc etc etc'. It went on like this for a few pages: things like, be mindful of such and such and call the support line if you need help. I'm not sure whose brainchild this book was, but seriously, if you know that a large proportion of your people are potentially coming home broken or even just different, maybe do a little more than just give a handbook to the wife so she can pick up the pieces! And just like that, life returned to normal, well so I thought. I guess deep down I knew there was no such thing as 'normal' anymore.

MATT'S PERSPECTIVE

It never ceases to surprise me just how ridiculous and conceited some of those people were, these so-called 'experts' creating the advice. I'd honestly never met anyone like a lot of these people before joining the Army. The silver lining to all of this was that the good ones were the supreme example of the professional soldier, and without getting all gushy about it, if it wasn't for my fellow people from the Lancers, and soldiers like Al Bungate, Tommy Nauvusolo, and Rob Marema, I don't know what would have happened. These people were always there, showing me what it meant to be senior NCOs and Warrant Officers. These men displayed outstanding soldierly qualities and I cannot thank them enough for the time they spent with us.

Matt went back to the 12th/16th as a full time reservist. What an oxymoron that is! But at least he was alive and well, on the outside anyway. The nightmares – he called them 'active dreams' – slowly went away. I think he was reliving things he went through while over there, things he wouldn't – or couldn't – talk to me about and things I knew better than to ask about. In most cases those 'active dreams' consisted of a lot of yelling from him and quick jerky movements, as though he was reacting to something with quick reflex actions. I learnt very quickly to simply watch and listen and make sure he didn't go anywhere, but to never, under any circumstances, try waking him mid re-enactment. I tried it once and got a fair head butt for my efforts. Poor Matt, once he realised what he had done he was so sorry and embarrassed, but I just brushed it off. He had enough to deal with. So, I learnt to just let him go through the motions, resettle and then we could all go back to sleep. He would wake up in the morning none the wiser and seemingly happy. But overall, he was different – not a bad different or a better different, just different. And he wasn't finished.

The Green Machine was well entrenched in Matt's blood by now. It had become far more than just a job; it was his career and his life and we all needed to find a way to be okay with that. He wanted it to be officially full-time now; he wanted in and whatever it took to get there, he was going to do it. Don't get me wrong, we talked a lot about the process, the impact and what it would mean for the girls and me. Matt was very aware of what Iraq did to me. Unfortunately, trying IVF at the same time was not my smartest move, but you live and learn right? What Matt didn't seem to understand was my fear of him not coming home; that's not how he processed it. It can't ever be how a soldier processes it. But it is how I, the wife sitting at home, processed it. That fear was all consuming; there was no easy way to alleviate it. It was real and it hurt. It's a lot to ask someone to endure, no matter how deeply you love each other. While Matt didn't get the pain, the fear and all the rest, he did appreciate that it was real for me, and that choosing this career path didn't just impact him; it affected all of us. So, we talked it through. He needed me to tell him it was okay to go, that I approved of him wanting to be the best full-time soldier he could be because without us, it wouldn't have worked. So I told him it was okay, we would be okay and long-term, everything would be fine. Now I just had to find a way to believe it.

MATT'S PERSPECTIVE

Kaz and I were discussing me submitting my paperwork to transfer to the regular Army. My mates, Sol and Spu, were looking to do the same. However, I got a sniff that there were contracts to B squadron 3/4 Cavalry and, on looking at their calendar for the next twelve months, I knew if there was any chance of getting another 'job', this would be it.

Matt had been home all of three months when, in August 2008, the call for a second deployment came. It was no great surprise, although it did come faster than I thought and hoped it would. The Army was once again calling for soldiers to step up for a twelve-month posting to B squadron 3/4 Cavalry in Townsville on CFTS (Continuous Full-Time Service) with the potential to deploy to Afghanistan. And yes, guess who was signing up! Matt found out about it while he was away on some course and when he came home I knew what was coming when he said, 'Babe, there's something I need to talk to you about…I know it's really soon after I have just come home, and it might not even happen, and it isn't until next year anyway.' How could I resist that hopeful look in his eyes? It's funny how they always justify the request before they even request it, isn't it? It's like, 'Wow that sports car is so economical and we would look amazing in it and it is the most comfortable car to drive... So I bought one!' I remember a soldier telling me, 'It's easier to ask for forgiveness than it is to ask for permission'. That was only in response to him purchasing a computer without talking to his wife first, but I reckon it doesn't really matter to them what the prize or goal is. Anyway, I thought it was pretty funny – and pertinent – at the time and it's stuck with me ever since.

Anyway, I finally got it out of Matt about the request to go to Townsville and, as always, I supported it. But I will admit, it wasn't an easy decision to let him go again, to know that I was going to have to hold my breath again and wait and stay strong enough to not let it break us. But asking him to choose didn't seem right either. What I really wanted to do was scream at him, 'No fucking way am I going through that again!' I wanted him not to ask me. I wanted him to be done. I wanted him to be a doctor or a lawyer; a profession that saw him come home every night and sit safely behind a desk in an office somewhere every day. I wanted to be selfish and say no. But that has never been me. I wanted him to achieve a career that fulfills him, I wanted him to know what success feels like and live it and breathe it and feel like he was providing for his family and the self-pride that makes most of us get up every day and get out there and have a go. And if achieving that meant we had to make sacrifices and deal with whatever hand that decision dealt us along the way, then so be it. I was not going to stand in the way of him living rather than just existing.

Panic struck when I realised there was an opportunity for the girls and me to go to Townsville with Matt. The thought of uprooting everything and taking the kids out of school and away from their friends was daunting. I told Matt it would be unlikely that we would move up there as a family and the girls and I would probably just stay in Armidale and wait for him to return. Besides, he wasn't guaranteed a 'job' in the sandpit, so we could always go to Townsville for a visit.

Suddenly the next realisation kicked in. What if Matt is away for a full year on CFTS again? Where would that leave me and our IVF?

We both still desperately wanted a baby. So, thanks to Matt and the amazing way he helped me get through my own healing process, we decided to go back and finish what we started. Besides, there were still frozen embryos in storage and they were our babies. That was always in the back of my mind. I just had to find the strength to face it all again. I told Matt I wanted him right beside me this time so we could walk the journey together. He agreed.

Between my travels for work and the many courses Matt had to attend, we spent months passing each other at the front door. But somehow we managed to find a window of opportunity in October to try again. We had two embryos left, so when the doctor gave us the option to implant both, saying it would increase our chances, we agreed. The thought of potentially having twins was so exciting. I wanted this so badly that I let my guard down and allowed myself to dream. While I knew Matt would almost certainly be going away to war again, it was okay because I would be pregnant and then I would have our baby, or maybe even two of them, and everything would be fine. I don't normally live with my head in the clouds, but I think I just wanted this so badly that I was grabbing onto every glimmer of hope.

This time when I made the trek down to Sydney I had Matt by my side. We completed the transfer and everything seemed to go like clockwork. Several days later my blood test confirmed the best possible news. I was pregnant! I didn't care if it was one or two babies. It could have been four and I would have been ecstatic. We were finally pregnant! In other great news, Matt was promoted to Corporal. How great life was; Matt was moving up in rank and we were having a baby (or maybe two)!

Time moved quickly and before we knew it, the time had come for the nine-week scan. We were so excited. We were about to hear our baby's heartbeat for the first time; maybe there would even be two sets of heartbeats! All the heartache, the loss, the fear and panic; in this moment, it was all worth it.

There was no heartbeat. I just lay there. I didn't want to hear it. I wanted a heartbeat. I just wanted the sonographer to keep searching. I was so desperate. 'Please don't stop looking,' I was begging him. These were our babies. They were loved and I wanted to meet them, even just one of them. I don't understand how I lost them. I did everything they told me to. Why was this happening to us, again? I wasn't a bad person and Matt and I had so much love for each other and we were going to be great parents. I wasn't being selfish wanting to give him his own child. I made a promise to him.

The cloud of sadness and guilt were all consuming. I still had to work and be a mum and a wife. I still had to put one foot in front of the other. But I had died a little inside that day. I cried for the little ones I was never going to meet. I cried for the promise I felt I was breaking to the man I loved more than life. And I cried for the journey I was walking through because I didn't understand why it had to keep hurting so much. Don't get me wrong, I fully understand that bad things happen and I also appreciate that there were so many worse things happening to others in the world, but this was *my* pain, right here right now, this was mine and it hurt. As if it wasn't bad enough to no longer be pregnant, I now had to book in for a curette because my dead babies weren't being miscarried by my own body. It was like the last cruel joke. The man I had promised children to was right by my side watching me fall apart, trying everything he could to support me. I felt like I had put him in this situation and yet I don't think I would have made it through without him. The guilt I felt was like a double-edged sword. Matt was away a lot for training, getting ready for his move to Townsville and I was facing the daunting reality that I was going to be separated from him yet again, this time for a full twelve months. I had been through so much on my own already and the thought of doing it all over again, albeit without the IVF this time, was just too much to bear.

MATT'S PERSPECTIVE

'I'm very sorry but there is no heartbeat, no sign of life,' the sonographer told us. The two embryos we had implanted had failed to thrive. Everything was looking good and they were growing, but somewhere out in the Universe someone forgot to start their little engines.

I watched as my wife's brave façade was cracking. I will always marvel at the strength of women; they are strong in ways we men cannot fathom – and all the raw emotion was starting to spill forth. My intuition had done me a disservice; I was so sure that this time was the one. Everything was a bit of a blur. There were so many emotions as well as so many things going on in our lives that it all just jumbled together. Ultimately you just feel useless. You see, men just want to solve problems. We want to be the hero of the day, ride in and save the day like they do in those old westerns. Unfortunately, this was not something I could fix. Only time could heal wounds like this.

One thing that was clear was that my wife was dying inside. To make matters worse, the babies would not abort by themselves meaning surgical intervention was urgently required. As a husband you feel super helpless and there is nothing you can do or say to fix it; you just have to be there, listen and try and hold it all together as best you can, support her as much as you can, take nothing personally in these moments, and provide comfort. But it's also important to let her know that you are affected by the situation at some stage. These are things I understand in hindsight. Like most things I tend to learn my biggest lessons the hard way, but I always try to make the lesson count.

Towards the end of November 2008, Matt was offered the CFTS contract to move to Townsville. The following January, the girls and I followed. In the end my decision was easy. I was not facing a full year alone. But the girls had to be on board too. We discussed it as a family before Matt left and they decided they wanted to go too. As the time to leave grew closer, they became more and more excited at the idea. Even if Matt did go to Afghanistan, at least I would have some time with him before he deployed.

MATT'S PERSPECTIVE

I remember the day the decision was made. Kaz and I were out in the backyard of our little place in Invergowrie, not long after losing the babies. She was upset. I gave her a hug and she looked up at me teary-eyed and said, 'Don't go'. I looked at her and suddenly the idea came to me. 'Let's all go to Townsville!' I didn't have to convince her too hard. It took a bit of extra organising, but it would mean we could keep trying and hopefully this time it would be less stressful for Kaz.

It's funny how things turn out. I was still broken inside, gutted at what was lost, but we finally made the decision that we were not going to have our own children and we would stop trying, as sad as it was. Just making that decision caused the clouds to slowly lift. The pressure of wanting something so badly that seemed so far out of reach had been doing me more harm than good. I don't think Matt ever truly thought we would not have kids. I'm pretty sure he never actually gave up. I just think that, to keep me sane, he set us down that path, telling me it was okay, and he was fine with whatever happened as long as he had me. I just needed to let go, enjoy the next twelve months and what would be would be.

MATT'S PERSPECTIVE

I have always been someone who listens to their intuition. I believe that if you break down any situation, your instincts will tell you which way to go. Unfortunately, it doesn't work at the casino, but it does work for most human related predicaments. It has kept me out of trouble for the most part and helped me judge the character of people. My intuition tells me if someone is ethical and trustworthy, both things that I find to be essential traits in those we welcome into our home. Similarly, I knew Kaz and I would have children, it was just something I knew deep in my gut. I cannot explain it any more than that. I just knew that at the right time and with the right schedule and a more settled life, we would get there in the end.

The end of 2008 gave us a chance to reset and heal, as well as get the kids ready for their new adventure and put together a farewell bash for all our friends who had become just like family in Armidale. We decided to rent out our home and move into Army housing in Townsville. This would give us a great financial advantage allowing us to push forward with building our little empire. We both knew we wanted to get ahead financially, and had worked out that investing in property was going to be the best way for us to achieve that. As we were both on pretty good incomes and we would be renting out our home in the new year, Matt decided we should look into buying another property before we left Armidale. He did so much research and, after looking into several options, he found a block of units for sale and decided that was the one. We organised finance, building and pest inspections, made an offer, read the contract, everything was in place. Then right at crunch time, I got cold feet and pulled out. I'm sure Matt was furious, but he didn't show it. He could see what a good bargain it was, especially given the potential. But he could also see what he was asking of me in his chosen career. I guess he just learnt to pick his battles. He decided it wasn't worth going forward if I wasn't completely comfortable. I wasn't, so we didn't. I'm sure he still thinks about those units and mentally kicks me every now and then. Oh well, back to the drawing board. I'll give Matt credit; he knows how to work the madness in my head. I'm sure he was not thrilled about losing the units, but I also know that he was certain he had planted the seed in my head and would get me across the line sooner or later.

Our farewell bash turned out to be massive; all the friends we had made in Armidale came along to wish us well. I had so many mixed emotions about leaving because I had made such good friends here, and they had become well and truly entrenched in our lives. I had been through so much with a few of them and they had helped me get through when Matt was away last time, more than they will ever know. Being with Matt meant more to me though. I felt confident that the new year would be the start of new beginnings for our family and new adventures. I was determined to make the most of the next twelve months and whatever our new life might bring.

CHAPTER 9

Here we go again

The girls were having some time with their dad before they moved to Townsville, so Matt and I decided that we would start the adventure with a road trip north. We decided to take a few days to drive to Townsville, stopping to check out some of the towns along the way. I had never been past Bundaberg and it had been so long between holidays for us that we decided to treat the trip as a mini break. It was a massive hike though. Between us we had two cars, a trailer and two dogs. Matt bought a set of walkie-talkies so we could communicate while on the road. We worked out that it would take us three to four days to get up there with long stretches of driving, so it was important that we could chat without the cost of phone charges. We bought mobile dog kennels and decked them out behind one of the cars and filled the other car with stuff we needed for the trip. It all seems so long ago now but I still have very fond memories of that road trip.

After a few days of drive, eat, sleep, repeat, I remember driving into Townsville and thinking, what the fuck? It was bucketing down! Not at all the image of paradise I had in my mind. We pulled over in the middle of town and devised our first plan of attack: find the dog kennel we had booked the dogs into. The Army puts you up in self-contained accommodation until you choose a house so we had to kennel them until we found a house. Once we got the dogs settled, we went to the unit, dropped the trailer and unloaded. The unit was basic but nice enough. We had one bedroom, a lounge area with kitchenette and a bathroom. The girls would have the adjoining room, which was open plan with a small bathroom. It was going to be just like being back in our tin shed, squashy but comfortable enough. Next stop, the shops to get some supplies. By this point it wasn't just bucketing down, it was raining cats and dogs with drops the size of tennis balls! By the time we got from the car into the shopping centre we were soaked from head to toe and sticky from the humidity. Welcome to the tropics!

I had taken two weeks annual leave to give me time to settle the girls in, find a house and find a good school for them. Townsville was new to both of us so we had to learn everything: where to shop, where the schools are, where to live and who makes the best coffee. Finding a house proved to be quite the challenge. Matt and I are so different in many aspects and our criteria for house hunting was a case in point. It didn't worry me if the house was a bit old and lacked style. I cared whether or not we had privacy and how easy it would be for the girls to get to and from school. Matt, on the other hand, really wanted modern features within the house and couldn't give a rat's arse about privacy. So, house after house after house, there was always something not right. By this stage we had been in the unit for a few weeks and I started to worry. You've probably figured out by now that I'm the worrier between the two of us while Matt just goes with the flow, always believing it will just happen. I started to stress about the dogs. The poor buggers had been left in the kennel all this time with only a few visits here and there from us. The girls were also getting ready to go to school and the school we had decided on was quite a distance from the unit, so it was all getting a bit much. Matt has a theory about me. He always says I lose it 5% before the end. Backing out last minute on the units in Armidale was a classic example. So here we were again. I had to travel back to Sydney for a week for work. I was feeling like shit. I wasn't enjoying the weather in Townsville; the relentless humidity even drove me to cut my hair short for the first time in my life. And we still hadn't found a house. As Matt had predicted, I ended up losing it and jumping on the next house we looked at. It was clean and modern, it had air conditioning (a must in Townsville) but...it was in a different suburb to the girls' new school, meaning I would have to drive them and pick them up daily until we found a bus route. To top it off, the house was positioned next door to a church-loving Bible bashing family on one side and Matt's sergeant and his wife on the other. So much for the privacy and safe walk to school I wanted! It all ended up being okay though because Chevvy (Matt's sergeant) and his wife, Tanya were cool. He and Matt hit it off straightaway and Tanya and I became good friends. There were other amazing WAGS in the street too and lots of kids. Even the Bible bashers learnt very early in the piece to keep their distance. Don't get me wrong; I have nothing against religion. I'm just not sure what I would call the 'higher power'. Buddha, God, Allah, Krishna, Confucius, to me, they (and all others) come from the same place. Stay humble at the top, hopeful at the bottom and help in between: that's my religion. But I would never force anyone to agree with me. Just respect that it's my view and if all others who follow any other faith could do the same, then we are all good. What makes me angry, and I have a very strong opinion on this, is when people who do follow a particular faith, stand in judgement of those who don't, like somehow they are a higher quality human being. Anyway, I digress.

We settled on the house and I flew to Sydney for work, leaving Matt and the girls to move us in. By the time I came came back, it was to a real home, dogs included. Only problem now was I still felt like shit. I felt like I had picked up some kind of virus when we arrived in Townsville and I just couldn't shake it. But I brushed it off assuming I was just adjusting to my new surroundings and coming off the back of a few pretty big years. I just needed everything to settle down so I could finally take some time to enjoy and appreciate life. I figured it would eventually pass and I would adjust. But I wasn't adjusting. In fact it seemed to be getting worse.

A few days later Matt turned to me and said, 'Babe, I think you're pregnant'. I laughed at him. Matt and his gut instinct again! Ignoring me he said, 'That's it, I'm off to the chemist'. And off he went. He returned about half an hour later and handed me a pregnancy test. 'Just do it. If I'm wrong, I'm really sorry, but babe, I'm not wrong'.

Well I'll be damned: two lines. Matt was right. We were pregnant! We fell pregnant without any needles, any extraction, any implant. Just two people who loved each other, who took a step back, breathed, relaxed and fell pregnant. Oh shit, Matt was right again. And he never let me forget it. To say I was elated is an understatement. Tears shed again, but this time for no other reason than pure joy. I wanted to wrap myself up in cotton wool. I wanted to climb into a cave and not come out for nine months. I so desperately wanted the nine months to just be over and our little bundle to be here safe and sound. But that was never going to happen and I was suddenly very aware that I would have to walk through some of this journey on my own. On top of that, I was experiencing morning sickness pretty badly. Turns out that's what my 'bug' was all along.

On the day of our nine-week scan I was a bundle of nerves, emotions still very raw from last time. But it was all very different this time. Finally! We have a heartbeat. We have a date. We have a viable pregnancy. And I'm making Matt a dad. It was the best feeling.

We settled into the routine of normal life, both working full time, kids at school, oh, and did I mention I was pregnant? The next few weeks seemed to fly by and before long it was time for the eighteen-week scan. I love the scans, I love seeing the baby grow and form and watching this little being moving around in my stomach. But life decided it was time to give us another reminder not to get too comfortable. Of course, it can't just be a normal pregnancy can it! I was advised to have an amniocentesis test, which comes with a risk of miscarriage. I didn't care at this point. If the risk was less than 1%, I'd seen this baby; I'd felt this baby moving. This baby was past the twelve-week risk zone. How do you get over that kind of loss? It becomes a burial by this stage, not a miscarriage. But I was urged to have the test, so I did. Thankfully Matt was there with me every step of the way. The Army may be tough but when family is in need, they're with you and the unit he was with was very supportive. We had the test done and I was told to take it easy for the next few days and look out for spotting or cramping. Once I was past the danger zone we had the agonising wait for the results that would tell us if I was carrying a baby with complications. Thankfully I wasn't. Our baby girl (yes we decided to find out) was 100% happy and healthy. We could breathe again.

As our house back home was rented out and we were paying reduced Army housing rent on dual incomes, Matt once again raised the idea of investing in property. All I could think of was, here we go again. Well this time things went smoothly and following a quick trip down to Brisbane, we became first time property investors after buying off the plan. Things were definitely looking up for our family. It was around this time also that we thought we would sit down with Jess and Ayleigh, now sixteen and eleven, so not little kids anymore, to ask if it would be okay if they could get used to referring to Uncle Matt as 'Dad' in front of their baby sister, especially once she was old enough to question why her sisters referred to him differently. Unbeknown to us though, the girls had already been talking because they approached us first. They wanted to know if we would mind if they started calling Matt 'Dad'. My heart filled with pride. I was thrilled beyond words that both my daughters felt so connected to Matt and intensely proud that they came up with the idea on their own. In that moment I think Matt was proud that they had come to this conclusion on their own too. It said a lot about his relationship with them and I was proud of him. He didn't have to step up and love those girls like he did. They gave him hell sometimes, as all kids do, but he bounced back every time and never once wavered. To this day, he has been a better, more stable, more resilient father to those two girls than both their biological fathers were and ever will be.

MATT'S PERSPECTIVE

I must admit this whole thing caught me a little off-guard. I had it in my head that the first one to call me Dad would be my little one. But I had been raising Jess and Ayleigh with their mum for the past seven years, since they were little kids. I picked them up when they fell and protected them like they were my own. They didn't deserve to have it thrown in their faces that they weren't my natural children. So I just went with it.

We were only in Townsville a few months when Matt got the confirmation that he would be deploying to Afghanistan in September.

We both knew it was coming but, to be honest, I was so busy with work, being a mum and caught up in the excitement of being pregnant and sharing that joy with Matt that I hadn't really given much thought to him leaving again in the last few months. But once we got the confirmation it slowly began to sink in, especially when I realised that it would mean he wouldn't be there for the birth of our daughter in October. The thought of having our baby in a strange town with no family or my closest friends around and Matt in another war zone scared the crap out of me. I had made some lovely friends in Townsville and the WAGS were very welcoming, so I was lucky in that respect. But that didn't take away from the fact that I was so far from home. So I made the decision that I was having the baby back in Armidale, our base, our home. To make it feasible, I needed to move back to Armidale with the girls by August. Poor Matt! Seven months into Townsville and one month before he is due to move to Afghanistan for eight months and he's packing up the house and moving his pregnant wife, daughters and three dogs (oh yeah, we bought a puppy in Townsville) back to Armidale. Poor Jess and Ayleigh too: talk about being bumped from pillar to post in a year! At least we were able to do it in one car this time.

Coming down from balmy Townsville to Armidale mid-winter when the average temperature is about five degrees was quite a shock. Matt was home for a few weeks to help me unpack and build the cot. If he wasn't going to be there for the birth, it was the least he could do. I don't think he minded building the cot, but as far as house moves go, by this stage I'm pretty sure Matt thought we would never move again. We also got to catch up with our amazing friends again and the Green Family put on a barbecue for Matt before he left. It was at the Sergeant's house, the guy who looked after me so well when Matt shipped out to Iraq, the one who granted us the unbelievable privilege of having our wedding reception at the barracks, the guy whose wife had become one of my closest friends. So I was home. I felt safe. Sad, but safe.

MATT'S PERSPECTIVE

Moving day was a royal pain in the arse. I had just completed a twenty-click stomp with the team I was deploying with, knocking off at 1 am. I drove back to the holiday apartment we were staying in, as we had handed back our house, leaving me just a precious few hours to scoff two quarter pounders, sleep and get back to work for any final tasks that needed to be done before picking up Kaz and hitting the road. We drove till 1 am the next day, caught a couple of hours' sleep in a motel, then on to Armidale. It was a very long, exhausting twenty-four hours.

On the twentieth of August 2009 a group of us accompanied Corporal Matthew John Page: husband, stepdad and soon-to-be new father to the airport to say goodbye and watch him fly out. I was seven months pregnant with our first-born child. One of the people who came along that day was our friend, Solomon Hanks, who at the time was an Army reservist like Matt. Soli had deployed with Matt to Iraq and the boys had become close mates. On the day Matt was leaving he said to Soli, 'Take care of my family while I'm away'. And Soli did just that. He checked on the girls and me almost every day. He organised a couple of the boys, mainly Spu (James) and Richo, to come to the house and mow the lawns. Sometimes he brought them out for dinner or just a quick catch-up. The three of them were amazing. I don't think they will ever realise how much I needed them throughout that period. They were connected to my husband on a level they didn't realise, all of them, but especially Soli. I just wanted to feel close to Matt, and those boys and the Army were all I had that connected me to him, them and this precious life growing inside me who wouldn't get to meet the man who brought her life until well into her first year. It's a bizarre sensation and I don't think I could ever really give the mixed bag of emotions I felt justice in words. So many WAGS reading this will be nodding their heads right now in agreement. They will just get it.

CHAPTER 10
Under the weight of it all

Communication with Matt was very sporadic so I never really knew if he was on base or 'outside the wire' (off base), other than when we had the chance to talk. When we did, he wasn't allowed to tell me anything and I wasn't allowed to ask. I had the computer on 24/7, just in case he came online. Facebook and FaceTime were our only reasonable means of communication as the phone line was constantly breaking up. I hated leaving the house just in case I missed him, so visits from people became my lifeline. Different time zones and war aren't exactly conducive to maintaining a healthy relationship, not to mention that I was still in full-time employment, working from home, and it was nothing for me to work forty, sometimes fifty-hour weeks and then be awake most nights, either struggling with pregnancy-induced nausea or just waiting, hoping Matt might come online.

The girls were great, aside from the usual teenage crap that went on, but nothing major. I had blown up like a puffer fish this pregnancy, which slowed me down quite a bit, so the girls helped around the house a little more than they normally would, and between the three of us, we made it work. There were no school runs to worry about because the bus pulled up at the front gate. Nonetheless I was exhausted, only I couldn't really see it; there was too much happening to be focused on me. Soli figured it out early though and realised I wasn't coping. He would sit quietly with me on the bad days and comforted me when I couldn't stop the tears. We were lucky to have him in our corner.

The constant niggling thoughts of 'What if…?' were wearing me down. The thought of having a baby that may never meet her father started to kick in, and trust me, it was a massive kick. What do I tell her if Matt dies? How do I explain that he went to war and he died so she would never get to meet him? Fuck, what if Matt actually dies? Have you ever been physically hurt so badly that it made you want to vomit? A relentless pain you can't escape, that feels like it's cutting you in two? That was what it felt like for me; a constant pain in my heart and head, every waking moment, every time my baby kicked or moved, or in the minutes of silence when I was left with nothing but my thoughts. I couldn't control this. I couldn't bring him home and make it all okay. And I didn't have an answer to, 'What if he dies?', not an acceptable one anyway. All I had was fear and uncertainty. I panicked every time I thought about how I was essentially creating this child all on my own while already raising two kids alone. I had to do my best every single day to hide the fear and panic from the girls. There was no point in three of us being afraid and anxious.

As the days rolled on, the intensity of my fear only grew stronger. The fear of him not coming home was bad enough the first time, but this time I also had a little person growing inside me that was his, that would need to meet him, to be raised by him, to know him as she grew up. Thoughts and questions of, 'He needs to be home. Why isn't he home? Why did he choose the Army? Why am I doing this alone?' consumed me. The cold hard reality of it all was that there was truth behind my fears. Soldiers *had* died. Vehicles exactly like the kind Matt was in were being blown up every day. There were fights with real guns and real bullets and soldiers had been shot and killed. There were even occasions when the very soldiers our boys were teaching and mentoring would turn on our boys and wound or kill them. The possibility that something bad could happen was very real. All the extra hormones racing through my body combined with the little communication we were able to have meant that I just couldn't break the cycle. I would write letters to him, wanting to know why he did this. Why did he choose this career? He should have been here with me. We had fought for five years to have this baby. Why wasn't he here? It got to the point where, rather than be thankful and happy to hear his voice when he was able to get through, I was just sad, lonely, scared and angry. I was angry that he had left us. I was sad that I was alone and, most of all, I was so scared of losing him. No one taught me how to not be afraid of him dying in war. Where was the damn Army now? Why wasn't the Army making this okay for me? Didn't they want his undivided attention? We all knew they were listening to every conversation we had. Couldn't they hear the stress, the fear and the concern in my voice? Oh that's right, he was just another soldier and I was just a civilian.

As the birth inched closer and closer, my fear grew bigger and bigger. I knew he could hear it in my voice. But what could he do? He was in hell, a war where he didn't know what the enemy looked like. Was it the small child walking past? Was it the lady holding what could either be a baby or a bomb? Instead of supporting him, I was adding to his nightmare. It was like a vicious circle; all I wanted was to hear his voice, to know he was okay. But every contact from him made me angry because he was okay and not home with me. God, I couldn't win! I knew I was making it hard for him but I was so angry with him. I should have said no, the minute I discovered I was pregnant. But then again why did *I* have to be the one to say no? Why wasn't our baby more important to him? Why didn't he say no? On and on it would go.

MATT'S PERSPECTIVE

So, with a heavily pregnant wife, two teenage stepdaughters who were giving their mum a somewhat difficult time, and the need to relocate them back to Armidale, why go through with the deployment? The answer is not a simple one, nor was there just one reason. Everything I had spent the last twelve months training and 'toeing the line' for culminated in this trip. I had spent so many months away from home during which I had completed countless hours of training, worked many late nights and had qualified in numerous skills additional to my role, that I felt like it would have all been wasted had I not gone. Furthermore, leaving would have deprived the team of qualifications the Army had given me. You see, the Army instils an unfathomable sense of responsibility within us, not only to the men you command, but also to the ones you stand with and those who lead you. My responsibility was to provide the capabilities I had been trained for to the team. Essentially, fulfilling those responsibilities becomes the whole point of your being and denying that urge wouldn't just be hard, it would be downright damaging to your psyche.

Professionally, this was an opportunity to do my job in the most dangerous, most active theatre on the face of the Earth. It was where our mettle and skills would really be tested, and like most of the lads, I needed to deploy. I needed to see if I could conduct myself with the skill and gumption of my forbears. You see, both of my great-grandfathers had served with distinction. They had both had field promotions, citations for bravery, numerous mentions in dispatches and were awarded with medals for their service. I remember being told stories of their actions by my grandparents when I was a child. I was told how my great-grandfathers had been gassed or shot; of the near misses and the shenanigans they would pull during periods of boredom or to break the tension. My favourite story was of the time one of my great-grandfathers led a sortie to relieve men pinned down by shell scrapes. He did so by grabbing a still hot mortar tube and moving continuously forward among the enemy despite the imminent risk to his own safety, effectively destroying the gun pits, which until then had been carving canyon-wide swathes as wave after wave of humanity swelled from the Allied trenches. This earned him a field promotion and a warrant as well as a military award. I felt an obligation to do them proud but knew they were very large shoes to fill. I also felt a sense of obligation to my country, as clichéd as it sounds. I felt, and to a degree still do, that we are very fortunate in Australia. We have the kind of fortune that some people in the world only dream of, and by and large people take it for granted. Don't get me wrong, I'm not the kind of blindly patriotic dickhead who walks around with an Aussie flag as a cape on Australia Day. I just believe in my country; I love the idea of what Australia could be. I feel like this country is a toddler and still finding its way in the world. But a country is not built in a generation and there is much work to be done. I believed, and still do, that it is the responsibility of every citizen to be of service and to contribute. This country, and all that it could be, is an idea bigger than one individual and worth far more. I thought at the time that the only way I could contribute was to put my money where my mouth was and offer the one thing that every citizen has, but only the uncommon few will ever be prepared to offer up, and that is a willingness to bleed for the country they believe in.

All of these reasons (and the pay, of course) led me down the path of continuing with the deployment, despite the impending birth of my daughter, my first child. A much wiser and more considered version of myself now knows that working to improve one's world can be done in so many more ways than what I ignorantly believed in the past. Kaz has taught me this. She has taught me that you can't fix everything. But you can make small incremental changes for people and encourage them to do the same. If only there were another 99,999 of her, we just might make a dent in solving the world's problems.

Before Matt left, he bought Willow (we chose our daughter's name before he left) a sheepskin rug. He wrote on the back of it how much he loved her and how special she was to him, signing it, 'love your Daddy'. He wrote that just in case he never got to meet the baby girl we worked so long and hard to create, because he also knew that the 'just in case' was real. We still have that sheepskin. It is a gentle reminder that love and life are sacred, and nothing is a given.

In the dead of night, after the girls had gone to sleep and I had forced myself to walk away from the computer just in case he came online and there was nothing left for me to do but go to bed and try to sleep, that was the hardest time of all. That's the time you're left alone with your thoughts and fears. That's when you don't have friends or family keeping you occupied or work or housework to distract you. In hindsight I wish I had been stronger. I wish I could have beaten the fear on my own and not dragged Matt into it, but I wasn't. Thank God he was though. Many a divorce in the Army is the result of deployment and the unbelievable pressure it puts on a marriage. We were one of the lucky couples, because not only did we survive, but it made us stronger. I'm not privy to every divorce in the Australian Defence Force, but I honestly believe a lot of them would have resulted from very similar circumstances. Maybe had there been a strong network of support, a safe place where the wife could have talked about her fears, just maybe those marriages would have survived. I would have loved to have spoken to another Army wife who was struggling or had struggled and come through, someone who could have told me that it was okay to be selfish or to vent my anger with so it wouldn't have been directed at Matt. I didn't want my friends to judge him, so I never said much to them. How many of those couples would still be together today had that kind of support been there?

The last few weeks of my pregnancy were hell, physically and emotionally. Thank God Jussy and Soli were there for me. It's funny how people draw conclusions. Jussy could be there night and day and no one would bat an eyelid. But Soli's presence caused people to whisper. Did they not know how comfortable I was with him around? Didn't they understand how safe I felt every time I saw Soli's car drive up the driveway. Sometimes he came for coffee, other times for the whole night. Why should that bother anyone? He knew what it was like where Matt was; he knew the impact to the family, to me; he just got it. He never judged, not once. He never once wavered in being the best mate to Matt and me. And anyone who thought there was anything wrong with what we were doing, well good luck to them. I wasn't losing Jussy as a support person and I sure as hell wasn't losing Solomon Hanks either!

As if the pressure I was under already wasn't enough, at thirty-six weeks pregnant I received a call out of the blue one night from the very first sergeant Matt ever had, the one who was there for me last deployment, the same man who was so generous at our wedding. Remember that amazing man? Well turns out he had been cheating on his wife, Karyn! He was calling to tell me Karyn was on her way over and asked me to be there for her. No sooner had I hung up the phone than Karyn turned up, so distraught she practically fell out of the car. I have no idea how I mustered the strength but I was out the door in a heartbeat, picking her up off the ground and walking her inside. She was broken, devastated. They had been together for years and they had three kids. What was he thinking? Nothing justifies an affair. If you're unhappy, leave. Leave your wife with some dignity. After all, you did once love her. So here I was, heavily pregnant, husband in a war zone and this man I used to respect and think the world of has just taken a great big shit on my life and destroyed his family as well. What next?

When I was pregnant with my two eldest I had to be medically induced with both of them because they had decided they weren't coming naturally. So I had made the decision early to have Willow induced too. At this stage I was the size of a house and I'd had enough: this kid needed out! Poor Jussy had already had to make the five-hour trek up and down from Sydney several times due to some false alarms. So I booked into the GP to ask him to support my case. My normal doctor was away and I got the new kid on the block. She was young, arrogant and had never had children. I tried to explain to her that after two kids, I actually had a clue. I knew my body. I knew my limitations, and I had reached mine. She refused. She told me I would have to go two weeks past the due date before she would support me being induced. I told her everything that was going on. I wasn't hysterical, but I was angry and sad. I didn't want to give this doctor the benefit of my tears but I couldn't stop them. Still she refused my request. I walked out of that room broken, disgusted and angry. Jussy was with me and she knew I wasn't taking no for an answer. I decided then and there that I would head to Sydney. I was going to make someone listen and help me get this baby out today. I rang the clinic in Armidale and I asked – probably actually demanded by this point – to speak with the obstetrician. He was a lovely guy and I had talked to him before about Matt and where he was. He could tell that it weighed heavily on me and told me at the time that if I ever needed to talk to someone to give him a call. Well I needed to talk now. I'm pretty sure I was hyperventilating by the time he got on the phone. Jussy was trying desperately to calm me down as I screamed down the phone about the incompetent doctor I had just seen and demanding that he fax a referral to Hornsby Hospital because I was on my way there now. He finally agreed to write me the referral, but only on one condition: that I call in and pick it up and that I have a cup of tea with him before I left and let him check me over. I agreed, so off to Armidale Hospital we went. If there is one thing I do consistently well, it's fight for what I know is right. And I was right. This baby needed to be born and God help anyone getting in my way right now.

I walked into the maternity ward and the obstetrician was there to greet me. After my outbursts and crying I must have looked a right mess. Anyway, he asked me to follow him straight into a birthing room. I asked him why and he simply smiled and said, 'So you can have your baby'. I could have hugged him! That's another person that to this day probably doesn't realise the massive part he played in keeping me going.

Jussy started to call everyone, including Soli, and put them on notice that I was having the baby. Soli arrived at the hospital as soon as he could. As the pregnant wife of a serving member on active duty, I had been given a special number that was only to be called when I went into labour or in an emergency (preferably a life-threatening one). That call would be passed on to Matt, if he was on base, for him to return the call as soon as he could. I held onto that number like it was my life support. Matt called back and I let him know that I was getting induced, that our baby was on her way. We had a little time before it all began. Matt needed to get a few things sorted so he could be on the phone with me when it all happened. I told him to hurry before saying goodbye. The girls were at school, so everything was in place for me to have this baby today.

The midwives got me all sorted on the bed and then the senior obstetrician came in. I can't even remember his name. You could tell he was here to placate the 'crazy woman' (my friendly obstetrician must have had to work miracles to get this guy across the line). The senior obstetrician then proceeded to explain everything to me about how the induction would work and that it could take several hours before anything happened, like I was either a newbie to labour or a moron without a clue! I kept trying to tell him I had two kids and knew the drill. But no, the medical profession always seems to know better. After breaking my waters and explaining to the midwives that he would be back to check on me after his morning tea break my first contractions kicked in before his hand hit the doorknob to leave the room. Thank God they had me hooked up to the monitor otherwise they wouldn't have believed me. The senior obstetrician stopped and walked back to the machine to check it. Jussy couldn't help herself and blurted out, 'Well, she did try to tell you'. I cracked up and the two midwives in the room both smirked. The obstetrician just looked at me, shook his head and turned to walk out. I yelled after him, 'Don't go too far, this is going to be quick'.

Remember how hard I had to fight for this moment? Remember the GP who thought she knew my body better than I did, the midwives who forced me to have to go to my GP for help because they didn't listen to me, the senior obstetrician who thought I was full of shit, right up until it happened, exactly as I said it would? Thanks, medical team! Thanks for taking the time to listen! Here's what took place next. My labour kicked in pretty quickly. Jussy was by my side and Soli was in the waiting room. I had the obstetrician (the really nice one) and two midwives in the room. Thankfully Matt had managed to call through on my mobile phone just in the nick of time. As my contractions intensified, I could hear Matt yelling over the phone, 'Come on, babe, you are doing really well, I'm so proud of you, babe'. Just as he said it, a third midwife walked into the room, heard the male voice calling me babe and looked over at the obstetrician puzzled. Jussy held up the phone and said nonchalantly, 'It's okay, it's her husband; he's in Afghanistan'.

MATT'S PERSPECTIVE

I remember that morning. It was about 0430 Afghan time. The occupants of my accommodation tent were abruptly awoken by a subaltern opening the door of the tent, whispering as loudly as he could, 'Cunts, cunts, is there a Corporal Page in this tent?'

'Sir,' I replied.

'Go call home. I just got a message from the watch keeper.'

'Sir', I replied again as I slowly started to get out of bed.

The temperature outside was minus 10 degrees Celsius and I was super tired. I had only just come back in under the wire and hadn't slept in days.'Pagey,' he said in the sternest officer voice he could muster. 'You awake?'

'Yes sir,' I responded, reaching for my boots and pistol belt.

I spent the next couple of hours on the phone repeating the same mantra, trying to think of things to say to my wife. I was told they were going to be a while, so I was able to have some breakfast and put some better clothes on.

They were telling me to push and push, and then all of a sudden, they were telling me to stop, don't push. Last time I was told to stop pushing, the cord was wrapped around Ayleigh's throat and she was turning blue, so I knew the words, stop pushing were not a good sign. Jussy could see the worried looks on the faces of the obstetrician and midwives and knew something was seriously wrong. I was panicking and asking her what's wrong. I was leaning on her; I couldn't see the doctor. She just kept stroking me and telling me it's okay and to hang in there. Meanwhile Matt was on the phone clueless because he couldn't actually hear any one of us through all the commotion. The midwife finally told Jussy what she needed me to do, after which Jussy proceeded to force me into a position that I doubt I will ever be able to get into again. Then the midwife told me to push with everything I had because the baby was stuck and they needed her out on the next contraction. I was struck with fear and had no idea what to think. What they didn't tell me was that they were hoping the baby would pass through without having to have her collarbone broken. The baby had her head out when they realised she had shoulder dysplasia. I did exactly as they told me to and pushed like there was no tomorrow and suddenly out she came. They cut the cord, cleaned, weighed and wrapped the baby and I dragged myself to the bed. Jussy brought the phone over and I told Matt that our baby girl was here. He told me he loved me, he loved her and that he had to go, but he would call me back very soon.

MATT'S PERSPECTIVE

I was at the office till about 0815 when I remember hearing the most alien cry. It sounds awful but I honestly thought she might have been deaf or had some kind of syndrome, however I put it down to the physical stressors of being in the womb too long. I was just able to tell my wife I loved her before Justine took the phone and told me everything is alright but they just have to get Kaz and the baby all cleaned up and settled. Unbeknown to me we still weren't out of the woods.

From the time they broke my waters to the time she was out was just a fraction over two hours. When the midwife took Willow over to the other side of the room, both she and the other midwife gasped in unison. Instant panic again! What the fuck is wrong now? I asked them what was wrong and one of the midwives laughed and said, 'Nothing is wrong. We just weighed your baby. Have a guess at her weight?' As both my previous daughters were eight pound seven and eight pound eight, I took a stab at eight pound nine. They both laughed and said, 'Ten pounds eleven ounces'. Fuck me, no wonder she had shoulder dysplasia! Oh, and remember that other 'medical professional' who wanted me to wait another two weeks?!

MATT'S PERSPECTIVE

I remember leaving the office and heading to our hangars where we would muster every morning. The lads were all standing around Ash, or Sergeant Ashby as he would more regularly be known. He was not only my boss, but he also helped me run top cover for the lads. I was the troop corporal. He was great to work for. The lads were giving me shit for being late as I was now twenty-five minutes past muster and Ash asked what time I called this to be attending the hangars, tongue in cheek of course. Ash wasn't the sort of bloke to punish you unless you really tried to be a fuckwit. He was the first person I told that I was a father and he shook my hand and gave me a big hug. At the end of Afghanistan, Ash bought an Eeyore plush toy and asked me to give it to Willow. She still has it today.

Ash was a troubled soul because early in his career he was involved in an APC (Armoured Personnel Carrier) accident which resulted in the death of another young soldier. John 'Ash' Ashby was found not to have been responsible for the death, however the Army being the Army just would never let John forget it. He was always treated as the ugly stepchild of the Army. I think his time in Afghanistan didn't help, but it wasn't anything to do with the enemy at hand, but rather the result of being fucked on by your leaders, the ones who are supposed to be responsible for your welfare. Sadly, we lost John in 2014. I guess he was acting out in those last couple of years and eventually it all became too much, and he killed himself. It's truly a heartfelt loss.

Willow Rose Page came into the world on October 19, 2009, all ten pounds eleven ounces of her. As I was lying on the bed cuddling this amazing chubby little ball of redness I instantly fell in love. However a sharp pain reminded me that I still had more work to do. Despite the contractions though, nothing was happening. For the next two hours they gave me more drugs to keep the contractions going in an effort to remove the placenta. By the end of the fourth hour I was physically and emotionally wrecked and this thing just was not coming out. So they decided I needed to go into surgery for an emergency curette.

Are you serious, really?' I thought. 'I have just given birth to a brand new, precious, desperately wanted baby and now you are telling me that in my super hyperventilated emotional state you want me to hand her over and be knocked out for surgery? And her dad isn't here. What if something goes wrong? Her dad is in a goddamned war zone. It's not like he can hop down to the nearest airport, jump on a plane and fly home?' I look back now and I get how over-cautious I was being, how super sensitive I was. But to be fair, I had to fight so hard to have her and to have the so-called 'medical professionals' listen to me when I begged for an induction while my daughter and placenta were over-cooking inside me. Had they actually listened to me then Willow wouldn't have been so big, the placenta wouldn't have been cooked to charcoal and super-glued to my insides, and I wouldn't have needed emergency surgery.

I reluctantly handed Willow to Jussy and told her and Soli, who had come in from the waiting room once I was decent, that under no circumstances were they to let her out of their sight, not for a second. If anything happened to me and I didn't come back, then they were to keep Willow with them until Matt arrived back in the country. They promised, and I went off to surgery. I know I'm a worrier at the best of times, but in a hyper sensitive environment my worry goes to a whole new level. After everything we had been through to bring Willow into the world and with Matt still in a place no normal person ever wants to go, I was worried. I knew that if anything did happen to either or both of us, my only saving grace was that I knew both Justine and Solomon would make a point of being there for Willow for the rest of her life and never let her forget how loved she was by her mum and dad. You couldn't wish for two better people than Jussy and Soli.

All went well with the surgery but the events of that day brought up many questions and thoughts that linger with me to this day. I often wonder if things might have turned out differently if partners of those serving in a war zone had the support of a Defence advocate, someone who understood how much of a toll the situation takes on those back home, someone I could have called and asked for help when I needed it. I don't know if that is the right solution but I do know that feeling like I was on my own with no one in the civilian world able to comprehend what I was going through, including the medical experts, was harder than I thought it ever would be. It made me understand why some wives call it quits. They aren't choosing to necessarily leave their husband; sadly many are simply leaving their husband's job. Unfortunately leaving the 'job' also means leaving the soldier they fell in love with and married.

After Willow was born there were so many mixed emotions running through me, from the excitement of finally holding our baby in my arms to the sadness that she hadn't yet met her dad and the fear of 'what if she never does'. Fortunately I had many distractions. In fact I had people coming and going all the time. Willow was the first grandchild on Matt's side, so his folks were very excited to come and meet her. I had friends from Townsville come by for a meet and greet and friends from town would pop in for coffee, so my days were easily filled. Soli was still amazing and there to give me a hand with whatever I needed, including killing a baby brown snake that had somehow managed to enter into our family room and was slithering straight for Willow's nursery. Never a dull moment in the Page household.

CHAPTER 11

Making precious memories

It was a strange shift and release of stress for me after Willow was born. She was finally here! We had our baby and I had delivered on the promise I had made to a man I would walk to the ends of the earth for at long last. I still desperately wanted him home with her – with us – but the intensity of my stress settled down as my focus switched to four-hourly feeds, nappy changes and bath time. Matt was due to come home in December for what the Army calls 'Relief Out of Country Leave' (ROCL) so I knew I just had to make it through until then, somehow, one foot in front of the other.

I filled the days by settling into a routine with this little bundle of joy (okay, not so little) and making sure Jess and Ayleigh stayed on track. That was starting to become more and more difficult, especially with Jess who was kicking into difficult teenager mode big time and giving me so much crap it wasn't funny. Ah, the joys of teenage daughters! Fortunately, between our Army families from all around Australia and our friends and family in Armidale, Sydney and the Central Coast, I had a fairly constant stream of visitors keeping me occupied through those first couple of months. I count my blessings for their company and support and for giving me some respite from dirty nappies, feeding and teenagers.

During those long weeks between Willow's birth and Matt's upcoming home visit to meet her, I did my best to make the most out of the limited communication I received from Matt. Whenever we could, we would FaceTime and I would have Willow on camera. It was all about Willow and Daddy now. I took photos every day and created album upon album to upload to Facebook so he could see all the precious moments I managed to capture.

MATT'S PERSPECTIVE

I was so grateful to Karen for posting those photos. She didn't have to, and I knew she was angry with me for being away when she needed me, and to a degree she was right. As husbands it's our job to help protect our loved ones. I remember the first image I saw of my little girl. Everyone thinks their baby is the most beautiful thing on earth, but not me. Willow was this ugly, wrinkly, humungous beast with a face like a squashed tomato, but I loved her all the same and with all my heart. Today it is an ongoing joke between me and Willow about how ugly she was as a baby.

Soli and the boys kept up with their regular visits, which I appreciated more than they will ever know. They would come out a few times a week to help out where they could and for a chat. They would ask about Willow and how she was settling in, chat about things going on in their world and take the piss out of each other, I'm pretty sure just to make me laugh. It worked and I was grateful for the distraction. Every so often Matt would make contact while the boys were with me so they would get to say hi. But they always made it super quick to give Matt and me a chance to catch up. I never once felt like I had to entertain them nor did I feel like I had visitors when they were around; they were family. Soli seemed to sense when I needed to get out of the house and would call from time to time to ask me to meet him in town for a coffee. Without him I'm sure I would have ended up a hermit. No matter what was going on in his own world, he was always there to look out for us.

Real friends are amazing and I can honestly say I have had some incredible friends come into my world who will be with me till death us do part, but the friendship I shared with Soli and the boys took the meaning of friendship to a whole new level. Not once did I feel as though any one of them wouldn't be there for us in a heartbeat. No matter what or where they were, if I needed help, one or all of them would be there. I even had a full load of them turn up in two trucks one day. Apparently they were driving through on the way to an exercise and just wanted to call in to say hi. That's one thing I do cherish about the Army. They don't just have each other's backs; it extends to the families too.

One night Matt organised a special dinner for the girls and me at our favourite restaurant in town. There were flowers, the meals were pre-ordered and everything had been planned to perfection, even to the point of Soli picking us up to drive us to the restaurant and return to take us back home again. It was pretty amazing how the boys worked together to make it happen. While Matt wasn't there physically, everything about that night was Matt to a T. Despite the distance between them, somehow he and the boys orchestrated everything to make it feel like I was having dinner with my husband that night, even if only in spirit.

The countdown was on. Matt was coming home, not for Christmas, but I didn't care; we were finally going to get a photo of Daddy and Willow together. The anticipation and excitement were intense. I was driven by a desperate need to have that one photo of the two of them together, a memory Willow could have later in life. I wanted him to feel her, to hold her, take her in and experience the outcome of our creation. He was flying into Sydney and I wasn't going to miss it for the world, so I decided to pack up baby Willow and drive the four hours to my parents' place a couple of days before he was due to arrive. I left the older two at home as they were still in school, and Matt would come home to the farm for the majority of his ten days' leave, so they would have time with him as well. I was so excited I just wanted it to be the day already. Here I was, once again building memories before they happened, as only I can. Only this time the actual memory would far outweigh anything I could have pre-empted.

The day finally arrived. I got up early to get everything ready, including bathing Willow to make sure she looked her best when meeting her daddy for the first time. Mum had this old washing tub, which we used in the kitchen to bath Willow, and it still makes me smile to picture her only just barely fitting into it. I have a photo of the tub sitting on the kitchen bench and Willow in it, with my dad shaking his head in amusement in the background.

To say I was excited would be an understatement. I felt like my feet weren't touching the ground. All I had left to do was feed and dress Willow in her prettiest outfit, well my non-girly version of 'pretty' anyway – a lovely yellow T-shirt all-in-one with a pair of blue denim overalls – pack the car and get my arse to the airport. Although I was several hours too early, I couldn't wait any longer. I didn't care how long I would have to wait. I would have been there several days early if I needed to. The date was December 12 and here I was at the Sydney Airport arrivals hall with our now eight-week old bundle of joy tucked snugly in the pram waiting to meet her daddy for the very first time.

Before Matt had left for Afghanistan, we bought our baby-to-be a sheepskin rug on the back of which Matt wrote a special father to daughter message, something she could cherish forever, no matter the outcome, but especially if the outcome meant losing him. We also bought her a cuddly toy that recorded a voice message. Every night for the past eight weeks I played that message for Willow when I put her to bed so she could fall asleep hearing her daddy's voice. But this was to be the first time she would be able to connect with him using all the senses; to feel his embrace, see him and breathe in his scent so she could truly begin to bond with him. It already felt like he had missed so much. Not today though. Today was the day I would finally get to take a photo of our daughter in her daddy's arms.

We waited and waited. Plane after plane arrived, none of them his. Then after what felt like an eternity, the status of Matt's flight from the Northern Territory, his first entry point into Australia, changed to 'Landed' on the Arrivals board. He was here! He must have been the first off the plane because next minute – well, so it seemed anyway – there he was, walking through the gates. Our eyes locked and we both smiled as he walked over to me. We hugged, no words needed, we just held each other in silence. Tears of joy streamed down my face. In that moment everything melted away: the anger, the questions, the fear, all the negative emotions were gone, replaced by relief and joy. Fear is such a powerful emotion and it can generate so much more. It can be a catalyst for anger, sadness, bewilderment, all kinds of emotions that are not easy to explain or control. But when the fear is gone, they all seem to melt away with it, maybe not forever, and certainly not in my case, but right there in that moment, all I felt was love, happiness and an overwhelming sense of relief.

Now it was her turn. 'Do you want to hold her?' I asked. I picked up our baby girl, quickly grabbed my camera and gently placed Willow in her daddy's arms, ready to capture the first moment my husband held his little girl. I don't think I will ever really be able to explain what that moment meant to me, getting that very first photo, and only very few people in this world will ever really get it. My heart breaks a little every time I read this part of our story, because I know there are mums out there who never got this moment. To each and every one of you, I am so grateful for your husband's service and deeply sorry for your loss. I am in awe of your ability, strength and courage to move forward and build a life for yourself and your family.

As strange as it may seem, even though I knew he had to go back to the sandpit, back to war, and the threat was still as real as ever, I had let go of my fear, at least for the time being, knowing we would have him for the next ten days. It would be a time to make precious memories and take loads of photos and videos that I would one day be able to give Willow to show her how much her daddy loved her should the worst ever happen. I was determined to make every second of those ten days count. I was going to record every single moment of it so she could see, at least on some level, the man he was to her.

In reading this I'm sure you can see that, while the fear had subsided, it was never far away. I was always conscious of the harsh reality of war and that our time together could be short and was therefore so very precious. I know that's the journey we agreed to walk and that's what memories are made of: the good, the bad and sometimes even the ugly and heartbreakingly sad. But that moment at the airport was ours; no one could take that away from us. Matt's luggage got tossed in the pram. There was no way he was letting go of his little one.

They bonded immediately. Although Willow was just eight weeks old, I had progressed her to combination feeding so Matt could feed her as well. I wanted him to experience as much as he could in those ten short days. I didn't really notice at first, but once the excitement of his arrival had settled and I took the time to get a good look at my husband, I could see that he had wasted away to almost nothing and that he looked so very tired, like he had hardly eaten or slept for months; in truth he probably hadn't. But he was just so grateful to be home with all of us, especially his baby girl. He didn't want to miss a thing and was determined to do everything. I'm pretty sure I could have slept the ten days away and neither of them would have noticed. He was bottle-feeding her, bathing her, burping her and changing her nappies. He loved every minute of it and she settled so beautifully in his arms. I have always said to him that the place I feel the safest is in his arms, and it seems Willow felt the same. For ten days and ten nights, she was his.

People often ask me how the hell I survived and tell me they couldn't do it. And to be honest, sometimes I look back and wonder the exact same thing. I wasn't the first wife to have a husband deploy, nor was I the first to have a baby while he was away, and I guess each of us copes in our own way. But I can tell you this much: without the support of Soli and the boys, Jussy, Jess and Ayleigh and a few very close friends in Armidale, I doubt I would have survived, I doubt this story would have been written and I seriously doubt my husband and I would still be together. I can see why so many defence families break down. I often wonder how many wives actually left their husband, as opposed to leaving the Army, still loving their husband but unable to cope with the lifestyle.

To this day I think about all the husbands and wives at home waiting and hoping for their loved ones to return safely on a daily basis. I think about all the men and women brave enough to stand up, put on the uniform and serve for their country, for all of us. I think of the fallen and the ones left behind, the children who didn't get to grow up with their dad, the wives who never got to be held again by their husbands. We read about it in the papers and see it on the news, but only for a fleeting moment before the news moves on to other stories. Yet those affected, those left behind, those still fighting and those still waiting, have to live and breathe their story every day. It's a hard journey to walk. When you take away all the bells and whistles and strip it down to raw emotion and endurance, it is a hard life for the soldier and a hard life for the partner at home. So every day those brave souls receive a silent minute of my time to simply say thank you. And every time I see a man or a woman in uniform – any uniform – I go out of my way to thank them because I know how much they are sacrificing to wear that uniform. My kids are so used to seeing me do this that they also now go out of their way to thank people in uniform.

MATT'S PERSPECTIVE

Those ten days were amazing even though Kaz and I had to work through some stuff. It was clear she still carried a lot of resentment towards me for being away at such a critical time. But as always we worked through it. As for Willow, well it was pretty cool. We took to each other like ducks to water. You can have younger siblings or cousins, you can be the best uncle in the world, you can raise stepchildren or foster children but you will never fully grasp the bond you feel until you experience your own baby. Everything is different. You change. It reaches inside you and pulls on something you never knew was there; the scent, the feeling, the weight of them in your arms, even things like the need to be in physical contact with your baby is something so primal it simply cannot be understood until you experience it. The best way I can describe it is that feeling you get as a kid on your birthday and you have to go to school. You spend your whole day with this underlying sense of excitement knowing there are all these amazing things at home waiting for you. Well, it's kind of like that only times ten. Don't get me wrong; I am not diminishing the role of step and foster parents; I am one myself after all. But it just feels so different.

Willow was a super chilled baby. I used to sling her over my shoulder and she would just just lie there, burp, fart and then hang her arms over my shoulders, let her head hang down and fall asleep. And that's where she would stay for as long as I would let her. I suppose you could say it was a time of contentment. I just focused on the time we had together. I knew that time was limited, but focusing on leaving again would only bring that day around faster and leave me lacking for the memories I would have otherwise made.

To be honest, it's all a bit of a blur now. I just remember it was as close to perfect as it could have been considering my imminent departure and the underlying resentment, anger, fear and a myriad of other emotions that were boiling just under the surface for my wife. I know that this time for Kaz was painful. While she was desperate to make the most of this situation, she was just barely holding it together. I don't think I fully understood or appreciated the total extent of what she was going through at the time and how much she sacrificed to make all of this happen. But as much as it was painful, it was equally important to her to make memories during those ten short days we had together.

By no means was this time completely without conflict. Kaz and I wound up in a few arguments during which I think I only scratched the surface of the underlying maelstrom of the situation. I don't think I will ever be able to repay the emotional debt to her. Kaz is probably the most selfless person I know. Today I think I have somewhat of an understanding and appreciation of her, but even that feels inadequate considering the depth of love and selflessness of this woman.

We spent two of Matt's ten days home travelling from my parents' home in Sydney to the Central Coast to see his parents and his grandparents. It was a significant milestone as Matt was the first grandchild and Willow was the first great grandchild. So many people wanted to catch up with Matt while he was home. When we finally made it back to Armidale, I organised a couple of dinner parties so we could catch up with a bunch of friends in one go. Between socialising we did our best to just chill with the five of us, but there was always someone excited and desperate to see him. These people were our family; they got me through. So they had every right to be a part of the ten days. After all, they were going to be the ones holding me up when he left again. Everything happened in such a whirlwind that we really didn't have time to think about how quickly it was all coming to an end. And then it did. Next thing I knew, my husband – the soldier – was packing his bag and the countdown began to take him back to the airport. We talked as he packed. We talked about the pain, the anger and the hurt. I know he tried his best to understand where I was coming from, but deep down I knew there was no chance in hell that he would. He had it in his head that he was exactly where he was supposed to be. He bled green and this was his duty. He knew I was hurting, and I know he was sorry for that, but I think to a degree he was lost because I had agreed to support his decision to deploy and now I was regretting that decision. To this day he has never really understood the fear I had about losing him. I get that because they don't allow themselves to ever think they aren't coming home. But at the same time, he didn't understand why I focused so heavily on the 'what ifs'. To be honest, I can't answer that either. I have no idea why it weighed so heavily on me. Perhaps how not to focus on the 'what if' scenarios is something that needs to be addressed under a support structure for defence partners. We talked through as much as we could in the time we had left, promising each other to try and just get through and not give up on each other. In the end we both agreed that we are both flawed and therefore perfectly imperfect together.

The dreaded D-Day arrived. It was Christmas Eve 2009. The car was packed and we were heading to the airport. The familiar gut-wrenching fear and panic kicked in and sat like a lead balloon in the pit of my stomach. Jess and Ayleigh, Soli and the boys and a few other close friends accompanied us to the airport. They all said their goodbyes and moved outside, giving Matt, Willow and me a few moments of privacy to say our own. He gave Willow one last cuddle and then turned to me. I just held him tight, told him I loved him more than life itself and begged him to stay safe, promising I would be here waiting for him to come home. And just like that, he was gone again. Soli was outside waiting for me. He gently took Willow from me, held me in his arms and let me release the tears I had been holding back all morning.

CHAPTER 12
Never a dull moment

Things at the 12th/16th unit had changed. There was a new guy in charge by the name of Danny, who was cool enough. But the changes, combined with the fact that Matt knew he couldn't sustain CFTS too much longer, made him realise that he needed to think carefully about his future and open up his options. So while he was in Afghanistan, he educated himself in personal training and rehabilitation. He figured that if the right opportunities didn't come along in the Army then he may as well aim for the next best thing in his eyes: fitness. He also planted the seed to go full-time Army and started talking about applying for SAS. Yep, there it is! He wanted full time, but only if he could get through selection. He definitely didn't want to be a 'grunt', the term they use for a normal front line soldier. My husband wanted to be the best of the best, Special Forces. We agreed to talk about it when he came home but I knew this was just the beginning.

I finally accepted that Matt needed to apply for SAS. We talked and I told him he had my support. I would do whatever he needed me to, under one condition: that we had a breather before the chaos started again. He agreed. He was due to come home in March.

Between Christmas and Matt's homecoming I took photos and videos of Willow every day to send him updates about the funny little things our two-month-old, our three-month-old and our four-month-old was up to. As the time drew nearer, I began to make plans for Willow and me to meet him, this time at Townsville airport where he was based. There was no way I wasn't going to be standing there with Willow waiting for her daddy to walk through those gates. Never in my wildest dreams did I imagine getting there would prove to be such a challenge. I have travelled all over world for work and could genuinely be called a seasoned traveller. How hard could this be? With no direct flights from Armidale to Townsville I had the choice between flying into Sydney and then transferring to a flight to Townsville or take the almost 1700 kilometre trek from Armidale to Townsville. I opted for the flight, not realising we would have a four-hour stopover in Brisbane. Willow was not a fan of flying and made the trip hell for everyone on both flights. I can honestly say I have nothing but empathy for parents of a screaming baby on every flight I have ever taken since. Willow and I arrived the day before Matt was due to fly in. This was finally it. Tomorrow he would be on home soil. I knew this wasn't the end, but this time I felt I was ready for the next chapter.

I thought we would have had more time once Matt was home. I knew things would go relatively quickly with SAS selection, once he applied, because they work to a strict timetable. But I guess I just hadn't thought it all through properly before he came home. Once back in Armidale, Matt settled into building a personal training business and before we knew it, he got the call he'd been waiting for. This call set in motion a hectic and chaotic twelve weeks during which he not only had to prepare himself physically for the gruelling SAS selection process in Perth but also build a PT business and spend time with his family, not to mention still being a Choc as well. Nothing like taking on the impossible and expecting it all to work out, but that was Matt, the eternal optimist. He just figured he would make it all work and everything would be fine. Because he wanted it so badly, he never stopped to question whether or not he was ready. He just dived in.

The fact that he had worked in fitness before leaving for Townsville had earned him a good reputation, so it wasn't hard for him to grow a clientele. Building a business strong enough to earn a decent living in case selection didn't come through was more than just working with a few clients on the side. He was still very green and being a Choc still trumped everything. If they needed him, they knew all they had to do was call and he would drop everything and come running. However, new times and a new chain of command meant there was less urgency for Choc soldiers to step up. And fewer Army days created more pressure to build the business. I wasn't worried from a financial perspective, not overly anyway. We were going to be okay. Matt had a plan in place and a backup plan as well. I was on paid maternity leave, so we were certainly not going to be destitute. We had money in the bank, a combination of deployment and savings. I had a solid career to go back to, we had the investment property paying for itself and Matt had a plan, whether it be in the SAS or private sector, to build wealth through property and enjoy life. I was okay with either as long as it meant financial stability and a life together that we were both enjoying.

In those twelve weeks Matt spent most days training clients, training himself and trying to spend as much time as possible with Willow, the girls and me. He would start before the crack of dawn, complete two client sessions before seven and then come home, fill his backpack with weights, put on his marching boots and walk or run for anything from thirty to sixty kilometres. Back home, feet blistered and bleeding, he would tape them up, conduct another client session, study for the SAS selection process and spend some time with us. Then off he'd go again, back to his clients before returning home once again to change into his uniform and head out to the barracks. This was our life, six days a week for twelve weeks. It was organised chaos but it was all for the greater good. He was on a mission and he wasn't letting anything stand in his way.

MATT'S PERSPECTIVE

I had completed my studies while in Afghanistan and talked to Kaz about the possibility of applying for SAS. Then, out of the blue, I received a call from a former boss who had successfully participated in an SAS selection course while I was overseas. To cut a long story short, somewhere along the line I must have made an impression on him because he was calling me to say that a lot of guys in his unit were currently on long-term or medical leave, which meant they were starting to feel the pinch. As a result they had more space and opportunities for anyone willing to attempt the selection course required for their unit. Great news! However, this was all I needed at this point as my life was settling into a groove and I had planned to put SAS selection off until the following year. But suddenly this bolt from the blue struck and fate played its hand. From that moment I couldn't stop thinking about the stories he told me of life over in the west.

I don't think I mentioned that call to Kaz for a couple of days. I'm unsure if it had slipped my mind or if I was deliberately withholding the information because I was supposed to be taking a year off for us to spend time getting to know each other again and enjoy raising our first baby together. I'm pretty sure it was the latter. But of course it was bound to come up in conversation at some point and when it did, Kaz actually reprimanded me for not biting the bullet and getting it done. She is an irrepressible woman and the pure truth is that I think she genuinely loves being the one to help you achieve your best more then she values her own successes, which are not insignificant. I cannot adequately express just how selfless she is.

I returned to work during this time, but I wasn't quite ready to go back full time. Fortunately CHEP was amazing. They accommodated my request for part-time work to see out the remaining six months. This meant I was keeping a wage coming in without sacrificing forty to fifty hours a week of my time with Willow. Although I was only part-time, I still needed to find someone to look after Willow while I was working. A mutual friend introduced us to a lovely lady by the name of Beverley McCrae who operated a family day care centre in her home just fifteen minutes' drive from us. Bev is one of those people who just radiate warmth from the moment you meet them; the type of person everyone can't help but love. She agreed to look after Willow and from that day on, Bev became known in our household as

Granny B. Willow adored her Granny B and her home and set-up were everything I grew up with. The children were allowed to get dirty, play in the mud and just enjoy being kids. She loved all the little ones that came to her house but developed a special bond with Willow, who became like a grandchild to her.

The twelve weeks flew by and before we knew it Matt was on a plane to Perth. He was gone... again. Neither of us really knew how long he would be away this time. All we knew was that the SAS selection process usually lasts several weeks, leaving those of us back home, including his PT clients, in a state of limbo. They were all super supportive though and and wished him well, promising to be there waiting for him should he return to the business. I take my hat off to Matt. Through those tough twelve weeks he managed to build a strong clientele, spend time with all of us and prepare himself physically for SAS selection, all with no down time for himself. Oh and we even managed to fall pregnant again at some point during those twelve weeks! Isn't it funny how we tried for five years, all the tests, injections, IVF fails and heartbreak only to fall pregnant naturally in the end, not once but now for a second time. And Willow was just six months old, meaning our babies would only be fourteen months apart!

I took Matt out for dinner not long before he was due to leave for Perth. I had wrapped the pregnancy test in tissue paper and tucked it into a jewellery box. Over dinner I handed him the box telling him I got him a little something. I was laughing so hard on the inside because the look on his face told the story that he wasn't happy that I had spent money on jewellery, something he had never been overly fussed on. But you could see he was holding back and knew he was going to have to suck it up and be nice about it, whatever it was. When he finally opened the box, the look on his face was priceless. You could see he wanted to be happy but wasn't sure if it was a joke. I told him it was for real and he leapt out of his seat to hug and kiss me. When he sat back down again he had a smug smile planted on his face, the kind of smile he has whenever he wants to say, 'I told you so', but has enough good sense to refrain.

MATT'S PERSPECTIVE

The news of Kaz's pregnancy with our second child was definitely unexpected, especially considering how long it took for Willow to come along. Now number two was coming and would just be fourteen months younger then Willow. Nonetheless we were both excited our family was growing. It was something for us to look forward to moving into the future. The news made me work even harder to ensure I was as prepared as I could be for selection.

I was finding my feet again. I was back at work part time, new baby on the way and working on myself in an attempt to be a better, stronger defence wife, because SAS was a whole new level and I had to be okay with it, all of it.

MATT'S PERSPECTIVE

'You have stress fractures in your left foot,' the doctor told me in a matter of fact tone from the other side of the desk. It was like a punch to the gut. I asked the doctor what that meant exactly and he responded that I would have to reduce load-bearing and restrict mobility. He also told me I would probably need to wear a boot.

Quickly processing this information, I made a last ditch attempt to change the outcome. 'Doc, you have to understand; I have a really important course that I need to be fit for. Can I do anything to get me through?'

The doctor didn't seem to understand in the least as he gave me no satisfactory answers. So I politely took the prescriptions he offered me and left so I could start forming a plan.

I told Kaz about the diagnosis and explained my well-considered and very mature plan that consisted of taping my foot until it could barely move, tying it into my boot and hoping for the best. Needless to say she wasn't happy about that but knew how much this meant to me so didn't say anything.

I spent the remainder of my time taping and icing my foot, which would swell up to three times its size during training sessions. In the end I had to scale back my workload so I could train and recover enough to get to selection.

I remember flying in to Perth with my mate Rooney. We came in from a regional area, so we had to fly in a day early. We had a quiet night and then it all began. People were rocking up all day and once it got to mid-afternoon, we were all assembled and the administration and testing began. However, it wasn't until the next evening that we experienced our first session of punishment. During our second set of knuckles-on-gravel push-ups, I glanced sideways to Rooney to see a grin on his face that matched my own. We were finally here!

Perth was a turning point for Matt. He knew he wanted it badly, but he also discovered that he couldn't do it all. It wasn't a pleasant discovery on any level. He had trained so hard physically to ensure he was ready for selection, but he hadn't really given the mental and emotional process much thought throughout those twelve gruelling weeks. He wanted it so badly he thought that alone would be enough. He never once stopped to ponder everything he had just been through and what all of that might actually mean to him when the dark thoughts came.

Left to his own thoughts and emotions while out bush on a survival section, all the effort and pain and blisters and time away from his family came down to that one section. It was in that moment that Matt finally realised the difference between physical preparation and mental preparation and the absolute need for both. He was feeling down and out, lost and confused and his head was racing a hundred miles an hour. So he made a decision. He would pull out, go home, spend time with his family, build the business and try again next year.

He knew he needed time, time he hadn't given himself between the sandpit and Perth. He needed to be Dad. He needed to make sure we were solid. I knew we were but at the end of the day, we had really only spent about five months together in two years. At some point during that period of solitude doubt had set in. He could have made it through, of that I'm sure, but he called it and came home. He wasn't done with selection; he just needed to do it right. He needed to rebuild himself and his dreams, only this time a little slower and with a little more control. While I was sad for him, I was happy to have him home. I was also really proud of him for having the fortitude to walk away. He was down, but he definitely wasn't out. Resilience has always been one of Matt's key strengths.

MATT'S PERSPECTIVE

That was a tough morning. I had stomped into the early hours and grabbed a little sleep in a re-entrant that had been washed out during the night. At some stage I got up and started to stomp again. I walked until I thought I should have been at the next checkpoint, but I just couldn't locate it. I sat down for breakfast and the dark thoughts started creeping in.

'You're shit, Matt. You're wasting everyone's time. Poor Willow, she deserves you being there to look after her'.

I radioed in and requested to leave the course. It was only after it was too late that I realised I was 100 metres off and didn't see it till the breeze blew the marker for the final checkpoint. The bitter irony of it all!

Matt's PT business was booming and his clientele grew rapidly through word of mouth. But he didn't give up on the Army and went back to being a dedicated soldier for the 12th/16th, making no secret of the fact that he would try again next year for SAS selection. He was only home a couple of months when whispers started rounding up again about another tour, another trip to the sandpit. Even though the PT business was going well, I knew there was no way in hell that Matt would miss out on an opportunity to go if the opportunity presented itself.

In November 2010 the 12th/16th went on a training exercise to Puckapunyal. I was seven months pregnant with our second child and Matt was part of the group going. It was a ten-day exercise away, which was pretty normal for us now. We had come to learn that this was just a standard part of the regular training for soldiers to ensure they were ready should they have to deploy. I was stronger at this point and we were also stronger together. We had spent time working on our relationship and making sure we would both be ready for the next round of selection when it came up. This was our journey now and and we were both equally on board this time. I loved this man and I knew his need to achieve, his determination to be the best. It is what makes him tick and what drives him a good portion of the time. And he was still very young; he still had a lot to learn about life. My role was simply to support him in his quest, just like he always supported me.

Little did we know that this was to be Matt's last ever training exercise for the Army and the beginning of the end of his military career. It was also to be a moment in time when I would lose the man I married and when the girls would lose the dad they knew and loved. It was the start of a new nightmare neither of us was prepared for.

CHAPTER 13
Welcome to Hell

It was November 21st, 2010. The girls and I were visiting my parents in Sydney when my phone rang. The number displaying was Matt's, which was strange because he rarely called me when he was out on exercise. You know that sinking feeling you get in the bottom of your stomach when it feels like something just isn't right? Well, in that split second my heavily pregnant belly sank.

I answered the phone and sighed with relief when I heard his voice. But my relief was short-lived because the next words he said were, 'Please don't get worried with what I'm about to tell you'. What the hell? Do people not realise that those words are like a red flag to a bull? When you pre-empt bad news like that of course you're going to worry! Especially a worrier like me!

Next thing I knew he was telling me he had been in an accident and that he was in hospital and that the accident had happened on the 19th. Yep, there it is; another red flag to a now raging bull.

'Okay, so hang on a sec, you're in hospital, because you were in an accident three days ago and you are only just telling me now? What the hell! Why wasn't I told sooner? Are you okay? What are they doing for you? What's happening?' I was vaguely aware of the rising tone and volume of my voice. Surely a courtesy call from someone to advise me that Matt was injured and update me on his condition was not too much to ask was it?

For the next few days all I had to hold onto were short and infrequent calls from my husband, not really telling me much and, as usual, holding back on the full-blown picture so as not to worry me. The part I struggle with here, besides the fact that my husband was lying in hospital away from home and withholding information from me to protect me; the part that really tips me over the edge is that the entire time he was in hospital, not one serving member bothered to contact me. Not one senior soldier, not one doctor or staff member at hospital, not one person at the training camp and not even someone from his own unit. Nobody bothered to contact me to let me know what was going on or to check in with me to see how I was coping with it all. And I don't care if they say it was at Matt's discretion to tell me as much or as little as he felt he needed to. Someone should have made a call to let me know, or at least check in on me. Surely that was not too much to ask?

It was several days before Matt and I were finally able to have a decent, uninterrupted conversation. It was only then that I learned what had happened. The group was out on a nighttime exercise. There were a number of Bushmasters on this particular exercise and Matt was Troop Sergeant. This meant he had to stand upright through the manhole of the Bushmaster leading the convoy. Just to put you in the picture, the Bushmaster is capable of carrying up to nine soldiers and all their equipment, fuel and supplies for up to three days. It's a massive armoured vehicle that weighs around 12,500 kilograms and is around seven metres long, two and half metres wide and more than two and a half metres high.

On the night of the accident, the convoy was heading out to train a group of junior officers, when the driver of Matt's vehicle accidentally drove the Bushmaster straight into a ditch, causing the vehicle, which was moving at a speed of around forty-five kilometres per hour, to come to an abrupt halt. Can you imagine the impact when a vehicle of that size and weight moving at speed stops so suddenly? Matt had half his body outside the vehicle, which was surrounded by the vehicle's Primary Weapons System. As the vehicle struck the ditch, Matt was thrown violently in a whiplash motion, the left side of his head taking the full force against the Primary Weapons System equipment to the point where it penetrated his helmet. Those helmets are designed to withstand the force of an assault rifle round, so you can imagine the damage he would have sustained had he not been wearing it. I would be writing a very different story had that been the case. Although he was disoriented and his legs wobbled beneath him, Matt told me he had to gather himself, control the situation and continue with the exercise.

MATT'S PERSPECTIVE

Next thing I remember, I was caught up in the gun ring by some of my equipment, which prevented me from falling. My driver, John, was freaking out, screaming at me, 'Pagey, Pagey are you okay? What do I do?' I told John I was okay and that he needed to head towards the feature far off in the distance where he would meet backup with the rest of the troop.

Later that night, as the convoy made its way back to the barracks after completing the exercise, Matt's condition deteriorated, including his cognitive ability. When the vehicles finally came to a stop, he was unable to disembark correctly. He lost the ability to use his legs and fell face first to the ground from the back of the Bushmaster. This was no minor fall. The back of the Bushmaster stands at just under a metre off the ground and Matt weighed around ninety kilograms fully loaded with all his gear. He was so disoriented that he couldn't remember not being able to use his legs and wasn't sure how he was able to get back onto his feet. His speech was slurred and, although he kept telling everyone he was okay, it was obvious that something was very wrong. So he was medivacked away from the site and taken to hospital. How nobody had the sense to lay him flat or put a neck brace on him is beyond me. His medivac from site consisted of throwing him in the back of a 4WD and driving him to the Army base hospital. Any neck or head injury he had sustained was now put at further risk due to the incompetence of the medics.

Why was my husband not medivacked with better care? Is it

because he was a reservist? Is it because he was on home soil? Or was it a lack of training of the medics? How many other incidents have happened to the detriment of the wounded on home soil? So many questions that will forever go unanswered because no one is ever held accountable. No one wants to answer or even own the hard questions. I was so angry, and I wanted to blame them, yell and scream at them, even burn them at the stake if I could. Why didn't they take better care of this man that had, like so many others, given so much? And why did he have to be so stubborn? Why didn't he acknowledge the seriousness of what was happening to him? The answer is that he couldn't; it wasn't the done thing. He couldn't and wouldn't let the others around him down.

The hardest part about men like Matthew is that they love you so much that all they ever really want to do is make you happy and protect you at all costs. And you can't help but love them for both those virtues. However, what they fail to consider is that we get to know them as men, their ins and outs, their good and bad habits, and we also can tell when there is more to a story than what they are telling us. By this stage, I could tell that my husband had sustained more damage in the accident than he was telling me. I could also tell that he was angrier about the way in which the accident had happened than he was letting on and that he was pissed off at the lack of care taken of him.

The next thing I heard Matt was heading back out into field after discharging himself from hospital because the doctors weren't doing anything for him. Because no one gave a damn, Matt then spent the rest of the training exercise out in the field, sucking down copious amounts of painkillers and anti-inflammatories to try dulling the pain and reduce the swelling to his neck, head and shoulders while having to deal with a constant flow of pink fluid from his nose. Who'd have thought that brain fluid discharging from your nose was okay! It was clear there was so much negligence going on. So many people had let him down. To this day I still can't tell you why because no one actually cared enough to find out. Not one investigation, not a single question raised. Not one person gave enough of a damn to protect the injured. Don't have a broken arm or leg? Didn't die? Then it seems you don't count!

The training exercise finished up and they were coming home. I figured I would hear more when he came home, but I also thought that, as I had not heard from him, he must be okay. I could not have been more wrong. Upon his arrival back to Armidale, I could tell that he was not right. It was like living with someone on a constant supply of low dose morphine. He was there but not really present. He would be talking about Willow one minute and all of a sudden forget her name; this child that we worked so hard to bring into our lives, this precious little bundle that he loved so much. And yet he stood in front of me and struggled to call her by name. He knew he was struggling, but because he had no idea why, he would become frustrated. Then he would lose his temper at his own frustration. It was like a vicious cycle he was desperately trying to break, and that quickly became his focus.

With each passing day Matt seemed to be in more pain, way more than he wanted to let on. I would see him quietly trying to down half a pack of Panadol, followed by Nurofen every single day. It was a never-ending cycle. Whenever I tried to broach the subject he would quickly shut it down. 'The Army doctor said there is nothing wrong so there is nothing I can do', is all I would get from him. He was becoming a very angry, constantly tired and frustrated man. He wasn't the Matt I was used to and I wasn't really sure how to cope with that.

Prior to the accident there wasn't a lot that rattled Matt. He would adopt a calm, controlled manner in almost any situation and his pain threshold could have easily been classed as high. But all of a sudden, this man I was married to had begun to display a level of fear I couldn't comprehend. He couldn't remember simple things; even to the point where he would go to say something and then question whether or not he had already said it. He would be so confused. Then he would become frustrated because he didn't know what was happening. He had constant headaches and sometimes, for no reason at all, he would just seem to disappear from a conversation. It was like one minute he was there with you, the next you were talking to a blank wall. This wasn't my husband anymore. There were snippets where I would get to see the man I knew, but they were becoming fewer and further between. I couldn't grasp the seriousness of the situation because the medical team at the Army hospital basically told him he was fine, so he went back to work. But he wasn't fine and as more time passed, the problems became increasingly obvious. The trouble was, you had to be looking, you had to know him, and you had to care in order to notice.

I didn't understand why Matt was turning into a complete stranger when the Army had cleared him. I couldn't see any damage on the surface, so why was he so angry and losing it with me for no reason? His display of anger towards me was never physical, but I did feel like I was constantly walking on eggshells. One day I could ask him a question and he would respond like the Matt I used to know, the next day I could ask the same question, even a simple question like, 'How are you feeling, babe?' and he would go off, yelling at me through gritted teeth, 'Why do you keep asking me this? I've told you I'm fine. What do you want from me?' I'll tell you why I kept asking. I kept asking because he was constantly holding his head, like he was suffering severe headaches all the time. I kept asking because I knew he was taking more Panadol and Nurofen than any human being should. And I kept asking because every other conversation I attempted to have with with him resulted in an increasing number of blank looks and silence. I would ask him if he realised he was having mini blackouts and he would tell me it was all in my head and he didn't know what I was talking about. But I also knew he was scared, as he had no idea how to fix this and could feel his sense of reality slowly slipping away.

There were times when Matt was completely fine, on the surface anyway. He could hide physical pain like no man I have ever met. It was really confusing for all of us at home because at times everything seemed perfectly normal, almost lulling us into a false sense of security thinking he was on the mend. Then bang, for seemingly no reason whatsoever – none I could put my finger on anyway – we were back to square one.

Matt was also starting to snap at the girls for the slightest misdemeanour. Where he would normally respond with a simple 'no' he would now release a tirade of ranting and raving. I would have to step in and tell him to take a breath, or just ask him to stop. His response would typically be to snap, walk away and then seemingly forget what had happened. It was like a switch was being turned on and off, only none of us knew where the damn switch was or who was controlling it. It certainly wasn't Matt! He would have PT clients booked in and be getting ready to go and then have to ask who he was going to train. He would be speaking to people one minute and black out the next. The blackouts would last just a second or two at first, so they were fairly easy for people to miss. However, over time they began to last longer and longer causing him to become frustrated, angry and scared as he tried desperately to hide them.

It got to the point where Matt could no longer train his clients. It was just too dangerous. But coming to terms with that was extremely difficult for him, this man who had spent his adult life thus far being in complete control of everything he touched, being able to lead and organise and build. He felt like he was losing it all and he couldn't explain why, to anyone, not even himself. I kept asking him to go back to the local Army doctor and seek further advice or get a second opinion. I suggested he ask his Sergeant for help. I just wanted him to do *something* and stop acting like nothing was wrong. But that would only cause more fights, more flare-ups, more anxiety for him and more confusion for me.

Eventually Matt had no choice but to fold the PT business and make the call with the Army. Essentially this entailed telling the Army they were wrong, that medically they had stuffed up. However, no one tells the Army they're wrong! The Army is a funny beast. A broken arm? 'No problem, we can put a cast on it and heal you'. A broken back? 'So sorry buddy, we will put you in a wheelchair, call you a hero and pension you out'. Please don't misunderstand me here. I am not underestimating the severity of any injury. I am merely trying to point out that the Army is quite proficient at looking after the physically broken. As long as the injury can be identified and proven, then the Army will look after the medical process for you. Die overseas and you are automatically a hero, and rightly so. The Army will be there to support your family, they will call you a national hero on plaques and in the media and sing your praise any chance they get. However, come to them with a head injury that they can't actually see, with symptoms that don't actually make sense and, well there just isn't a lot they can do for you. Oh, and by the way, if you're just a Choc who was deemed by the medicos to be fine, well then there is no support provided whatsoever. Welcome to our new world.

MATT'S PERSPECTIVE

The fact is, the Army did have an obligation. As soon as I was injured, I should have been medically downgraded, and my employment retained until I was fit again. However, they had an Army doctor sign me off as fit for duty. Fine, but here's the kicker: the chief clerk, aka 'chief fuckwit', decided to pull me off my promotion courses because, apparently, I was injured. Making sense yet?

I went to see my Training Warrant Officer to ask him for help and he said he would see what he could do. Next minute he asked me to sign a document. It was a record of training for night vision equipment. I told him that I couldn't sign it because I had never officially been trained on the equipment despite having used it on numerous courses and during various operations. He simply responded with, 'Well, if you could help me get my filing up to date then I will get onto chasing up what you need'. Such a nice guy...

The Army put Matt back on the books for seven days while they 'assessed' the situation. The Medical Officer at the time talked to Matt and deemed him fit for work. My God, the pain he was in, the fogginess, the outbursts of anger and the blackouts! Clearly Matt was anything but fine. But the problem was, the memory lapses came and went, he wouldn't always seem in pain, and he didn't appear to be foggy twenty-four hours a day. But I knew my husband and could see he was getting worse. I'm sure Matt didn't help himself either in the early days. He didn't understand what was going on and often didn't recall the blackouts. When he was asked by senior members of the Army how he was, he would almost always try and play it down a little, I guess so as not to appear weak. Well done, Army. Way to teach the boys! You break them down, you turn them into fighting, killing machines. But you don't support them when it goes wrong. Why was it so hard to see through his bravado and realise something was wrong? Why was it not a massive red flag when this man came to you after two tours of duty and a fast-tracked career of promotions? Did you not realise just how much it took for this man to come to you in the first place and ask for help? All you did was make it worse. You gave zero aid; you closed your eyes and turned your back on a soldier. So much for never leaving one behind. You all failed; you all left Matthew Page behind.

Matt went through hell during December. He was bounced from pillar to post by the Army's so-called experts and was never really taken seriously. Every knockback, every medical expert's seed of doubt, increased his own self-doubt, which intensified his downward spiral. The harder they made it for him, the worse the blackouts would become, which in turn increased his stress and anger. Shutting down his PT business mid-December was like the nail in the coffin. It meant he was losing his ability to provide for his family and losing who he was, all at the hands of a beast he didn't know or understand. Meanwhile the Army was non-committal about whether or not they would have him on the books or do anything to help him medically. They simply didn't want to know; it was in the too hard basket.

We were heading into Christmas, a time that is supposed to be full of joy and happiness, and I was about to give birth to our second child. At least Matt was going to be home with me as I brought this child into the world. It was meant to be our time, our turn to create beautiful, sacred memories together as a family. But it was all being ripped away by a monster I didn't ask for or understand.

CHAPTER 14
Keeping it all together

It was the catalyst of a disaster that never should have been. The accident was an accident, and neither Matt nor I are bitter or angry about that. However, had the Army taken the accident and Matt's condition seriously, had they not failed in their duty of care, then the journey to recovery would have already begun.

It was supposed to be such a beautiful time; we were about to have our second child. However, it was turning into an absolute nightmare, and in the middle of all of this, Ayleigh decided to add to the already volatile situation. Through all the upheaval and chaos, she had started to act out. I'm sure she was struggling with all the changes going on in our household, from moving back to Armidale and Willow taking so much of our attention, to starting at a new school and her mum being pregnant again.

I could shelter Willow from most of the upheaval. She was too young to understand what was going on and feel the intensity of it all. However, unfortunately the same couldn't be said for our second baby, Scarlet, who copped the full brunt while growing in my belly. I know she could feel it every time my stress levels rose. Jess, being a little older than Ayleigh, seemed to ride with it a lot better than her sister did. She could see the pressure I was under and did what she could to be there for me, for which I was so grateful. Ayleigh was too young to really get it and so, when she was given the chance to go to her biological father's place on the Central Coast over Christmas and the summer school holidays, she jumped at it. What I didn't realise at the time was that she was about to jump ship entirely. Through conversations with her dad, Ayleigh had apparently described being at home with us as a living hell. I didn't agree with some of the things Ken, my ex, had explained to me that Ayleigh had told him, but then there were some things I honestly couldn't dispute. Tensions were high in the household, times were hard – financially and emotionally – and on top of it all, I was ready to drop with my fourth child.

So, just like that, Ken decided that Ayleigh should stay with him permanently! No discussion with me, no regard for us, just an abrupt phone call to say she wasn't coming back and that he would come and get her things. I still remember Ken's words to me on the phone that day: 'Have her stuff packed up and ready. I have sought legal advice and there isn't a damn thing you can do about it. And by the way, I have been waiting ten years to do this to you'. I was gutted. I felt like I was not only losing a piece of Matt on a daily basis, but now I was also losing a child. What happened? Where did I go so terribly wrong? How could I feel anything but guilt and anger for letting my child down? And how the hell could I pull it all back together again? In that moment, my world crumbled. I hurt so badly. I couldn't control my sadness, my fear and my anger. For days on end I just cried. I couldn't stop the pain. My friends rallied around me and tried to help as best they could. But honestly, there was nothing anyone could do to stop this train from running.

Matt and I tried to talk to Ayleigh to see if we could get her to reconsider but Ken was on a mission and had made it sound like life on the Central Coast with him was going to be better than ever. I felt like I was losing her, which made me angry, but my hands were tied. If I called bullshit on some of the stories she was telling him, I looked like an arse. If I pandered to her and she came back, it would be a lie. In hindsight – and don't you just love hindsight – I should have just let her go without a fuss. Ken was her biological father and he had every right to have custody of her. And she had every right to go.

The battle I was facing with Matt was another matter altogether. It was massive and it took everything we had and more to get through. But at the time, I couldn't see logic because my mind was clouded by the pain. I wasn't angry at Ayleigh; I was angry at the whole world.

Most of all I was angry at myself because I couldn't figure out how to fix it for everyone like I always did. I eventually came to the realisation that I had to protect myself and baby Scarlet. So, I told Ayleigh I loved her and wished her the very best. I hoped for her that it all worked out and that the experience would be everything she wanted it to be. And then I walked away. She was thirteen and, for now, I had lost her. Her father hated me, and I knew he was going to feed her anger towards me and there was absolutely nothing I could do to stop him. I had no choice but to let it go.

Matt was angry that all of this was happening to our family, especially as it was slowly destroying us. At the same time, losing a part of him every day was adding significantly to the household stress levels. He couldn't work because he knew the combination of fogginess and short-term memory loss was dangerous to his clients, but he didn't see the impact it was having on the rest of us. Not only was he struggling with his medical condition and the lack of support from the Army to help him figure it out, but he was also having to watch me break over his condition and Ayleigh leaving. He felt helpless and it showed. He was like a caged lion and it was a dangerous situation because he knew he could no longer control the beast within.

We were all doing everything we could just to keep it together and put one foot in front of the other. Jess was amazing at this point. She just kept telling me it was okay, that I wasn't a bad mum and that Ayleigh would come around. At just eighteen, Jess was mature well beyond her years. She even told me that she believed Ken would eventually trip up and show his true colours, so it wouldn't be long before Ayleigh was home. However, we had made it clear to Ayleigh that once she made the decision to stay, she would have to stay put for at least a year. Harsh as it may sound, she needed to learn to live by her choices. We would soon learn just how right Jess was.

Scarlet Jae Page made her way into the world on January 4th, 2011, all eight pounds, nine ounces of her. I was diagnosed with gestational diabetes during my pregnancy with her so Matt and I worked together to keep it in check with diet and exercise. The strategy paid off. No more ten-pound babies! Don't ever let a medical expert tell you that it is an absolute necessity to have insulin. I tried it for a few days, but it sent me into permanent sleep mode, so I stopped and focused on diet and exercise instead. I'm not a doctor and this is not medical advice, just common sense. In my humble opinion, and based on my own personal experience, if it doesn't feel right, push harder for a better solution.

MATT'S PERSPECTIVE

Scarley Girl came hurtling into the world within less than an hour. I almost missed her arrival. Kaz told me to go home and have something to eat. I remember her saying, 'By the time you get back to the hospital I'll be good to go'. However, before I had even walked in the door, Kaz had called my mum, who was at our place, and said they had broken her waters and that it was time to have the baby. I waltzed straight into the kitchen and opened the fridge. Mum asked what I was doing, and I told her I was making a sandwich. 'Don't worry, Mum. The baby won't come before I finish my sandwich.' Mum was looking at me like I had lost my senses. Well, by the time I walked back into the delivery room I could tell straightaway that the birth wasn't far off. Within a matter of minutes Scarley came out looking like a skinned rabbit. She was as long as Willow had been at birth but, because of the way we had managed Karen's diabetes, she was long and lean. Even today the kid is all hard muscle and never gets fat, despite being the only person I know who can come close to eating as much cake as I can when I'm on a food bender.

Scarlet was perfect and beautiful in every way. And she also seemed very intuitive in that she appeared to sense everything that was going on around her. From day one she wouldn't settle with Matt. To this day he tells people that Scarlet hated him in those early days. My theory is that she could sense the stress and anxiety within him and within the household, which made her unsettled. We brought her home and it should have been awesome. But it was hard and sad and confusing. And she seemed to sense every bit of it.

After everything that had happened with Ayleigh, the way Matt was feeling and the lack of funds now coming into the home, Matt and I decided it was time to have a serious talk. He was finally beginning to accept that something was seriously wrong. We both agreed that he needed to take far more precautions and that we should start studying his patterns to see if we could find the triggers to his blackouts. Because they happened so infrequently and didn't last long, they were exceptionally hard to track. On top of that, he still didn't realise when he was having one. It became like a game of cat and mouse and I was the cat on the hunt. I had to watch him closely and monitor his every move.

Over the next few months, we kept going back to the 12th/16th, Matt ever hopeful that they would listen. He had so much respect for this unit and truly believed they would eventually see the error of their ways and help him. However, I could see straight through their bullshit. Management had changed, and it was quite clear to me that they were just trying to duck-shove my husband and he was getting worse by the day because of it. He became increasingly negative about everything and his negativity was starting to creep into every aspect of his life. He started believing none of this was real and he was making it all up like a hypochondriac. He was losing control and he knew it, and it frightened the shit out of him. It was like he was strapped to a rollercoaster that never stopped and he couldn't get off. The fear made him angry and because he had no control over his anger, we all felt like we were constantly walking on eggshells, not knowing what would trigger him off next. It became an endless battle. Every time I saw a glimmer of hope in him and I thought he was ready to fight, management at the 12th/16th would do or say something that would knock him back another ten steps. I felt like I was fighting them and his condition while simultaneously trying to save him, our marriage and our family, and keep our heads above water financially. It wasn't all day every day. There were moments when we just enjoyed the babies, a chat or simply hanging out. But those moments were few and far between, and the constant struggle to navigate his moods was starting to take their toll on me.

We also had to face the fact that Matt couldn't work, certainly not in his current state. I was going to have to go back to work eventually and my work meant I had to travel. God, what would happen then? What if we didn't have it all sorted by then? What would happen to him? I wasn't worried about the girls because I knew I could rely on Matt's mum. She would often come to Armidale to help us out and that was my lifeline. And if I had to go overseas, I could contact Cherie and ask her to come babysit while I was away. So at least I had that covered.

I don't think Matt's mum and dad really understood or even believed the depth of their son's injury at this point. I'm sure it was difficult to comprehend; it was for us too. In fact, at this point there weren't many people who knew the seriousness of the situation. Our circle of friends and family knew us as being very private and resilient people who rarely asked for help. It didn't occur to us to be any different now. So, I get it. I understand how someone on the outside would struggle to comprehend how much trouble he was in, especially also given the fact that there were no visible signs of damage. But what I will never understand is the Army's take on it. Everyone at the local unit knew Matt. They knew he wasn't the same man he was before the accident. And if the decision makers didn't know him personally, all they had to do was look at his impeccable record. The story spoke for itself. He had never complained, he had never asked for special treatment and he had NEVER asked for help, until now. They just didn't seem to care enough to make it a priority. It was becoming a common theme. If we can't see it, then it mustn't be real. And maybe Matt didn't know how to ask for help. Let's face it, the Army doesn't build soldiers to be weak. So, shouldn't his sudden pleas for help have raised the alarm, at least enough for them to investigate what was going on?

CHAPTER 15

Time to fight

Matt and I started discussing our options. We had savings, we still had the investment property and we were well ahead on our mortgage. However, assets and equity don't pay the bills and put food on the table. And Matt needed to do something. If he was home all day every day his condition, whatever the fuck this was, would most definitely kill him, and by now it was blatantly clear the Army was not going to step up and help. It was up to me to figure out what to do. Fortunately, I was still able to work from home. CHEP was amazing. As long as the work got done, no matter how and when I did it, everyone was happy. Even if it meant I had to be awake twenty hours a day, I would get the work done. Nothing was going to get in my way of looking after my family.

Unfortunately, Matt's options to keep himself occupied and produce an income were fairly limited. To be perfectly honest, the income was secondary. We desperately needed something more important than money right now; we needed a place he could go, a place where he could be safe and feel like he was contributing. A large proportion of Matt's Army training was centred on being a part of something bigger than yourself, to contribute, protect and serve. This was so instilled in him that the need to contribute was – and still is – a massive part of this man. Take that away from him and you take away his essence. We desperately needed to find something that would give him some of that essence back, that would give him a sense of purpose and something to focus on through the toughest fight of his life. I needed to know he was safe when I wasn't with him and I needed to know I could trust the people around him to look out for him and contact me immediately if something wasn't right. So, for all those reasons, combined with the knowledge that this decision would give us control and allow us to employ people we knew we could trust and rely on, we bought a coffee shop in town. Here we were, two people with zero small business or hospitality experience, and we bought into a Gloria Jeans coffee shop franchise. How hard could it be?

It took a while to get everything sorted. Matt had to undergo a six-week training course with Gloria Jeans at their Sydney head office. He stayed with his parents and it was good for him to get off the mountain and away from the reminders of everything he had lost. But it was also an interesting time because he still wasn't any closer to working out the cause of his blackouts, the fogginess and short-term memory issues weren't improving and, to top it all off, the medication he was on was barely taking the edge off the pain he was in. Despite it all, he did his best to hide it all from everyone.

Whenever we talked about the Army, Matt would visibly stiffen and become agitated and angry, knowing full well his career was done. Any chance of going back to Perth for SAS selection now was well and truly gone. If I dared suggest that he shouldn't drive, it would result in an instant fight and him telling me he was fine and I needed to leave him be. I found the calmer I could keep him and the happier the environment, the fewer blackouts he would have, so it was important to keep up the façade and let him feel like everything was going to be great. He didn't need the Army because this would be his next chapter and it would be fantastic. We both knew it was all bullshit, but I just couldn't be another source of pain for him and, right now, that seemed more important than anything.

During this time, I bounced between the Central Coast and Armidale, babies in tow. Fortunately, Jess was old enough to stay home, as friends rallied around us to make sure she was okay. As we got closer to opening the coffee shop, the fights revved up again and at times they were relentless. It wasn't really him fighting me, but it took me a while to realise this. He was fighting the anger, the pain and the loss. I just happened to be the one in front of him; someone to vent his frustrations on, someone with whom he could let down his guard. It was like a pressure valve; you just never knew when it would blow.

We could be discussing our action plan for fighting the Army when, all of a sudden, he would tense up and explode, yelling something like, 'They're not taking anything else from me, especially my freedom. They have already destroyed my career. Isn't that enough?' I felt helpless and couldn't find the words to respond. How could I respond when I knew he was right and there simply were no words to make it better? Through sheer negligence, they had destroyed his career and reduced him to a shell of the person he used to be.

So many unanswered questions, so much pain – physical and mental – and Matt knew he was hurting us too; he knew we were all struggling. I have no doubt there were moments when he just wanted to give up, but I could never bring myself to ask him not to do the unthinkable. It was as though saying those words would take it to a different level that none of us could ever come back from. Every day the fear was there, that he might just do it if the fight became too hard, the pain too much to bear. All I could do was hope that we were enough to get him through until I could find the right help. But I was way out of my depth. It felt like I was constantly juggling, running left and right trying to keep all the balls in the air without them crashing to the ground.

MATT'S PERSPECTIVE

Only a few people understand exactly what it feels like to wake up every day thinking that as soon as you get the energy to do it you will kill yourself. It's like a predator, always sitting off in your peripheries, just waiting to sink its teeth into you. It never really goes away. It's always waiting for just one slip-up. Forget to take your meds once, get yourself too worn down, and there it is, it's on you.

Before we bought the coffee shop, we had been talking to one of our close friends, Tash, who was working as a barista in Armidale and had been in the industry for a long time. She knew all the facets of running a cafe and was willing to come across and work with us when we took over the shop. She would help Matt learn the ropes and, importantly, would be my eyes and ears. Tash was one of my best friends and one of the few people who knew almost everything that was going on, so I trusted her implicitly and she knew exactly what to look out for.

By the end of April we owned the shop and Matt had completed the training. I talked to Tash and our key staff about the need to be vigilant and aware. I taught them what to look out for, how to manage him if he was down and how to manage the blackouts. Things went well for the most part, but it definitely wasn't all smooth sailing. I remember Tash telling me about a time when Matt was taking a customer order when suddenly, for no reason at all, he blacked out. Poor Tash. Realising what was going, she jumped in, gently turned Matt away from the cash register and, without missing a beat, took over the order. I'm sure the customer was as confused as all hell at that point, but Tash stepped up and just treated it as part and parcel of her day. The staff at the coffee shop and I could tell stories like that all day. We got so used to things happening, we would just end up shrugging it off and saying, 'Oh, that's just Matt'.

Barely two months in to owning the shop and I was worn out and at my wits end. I had two babies under two, I was back working and travelling, and I desperately needed to know Matt was safe. The coffee shop allowed me to control his working conditions and employ the right people to help him. But the business was draining us financially; everything we had saved, everything we had done to get ahead was on the line. Despite this however, that coffee shop would forever be worth more to me than any dollar value, simply because it gave my husband a place to go, a place to be safe and feel like he was a part of something. That place – more importantly, the people in that place, including some incredible customers – gave him hope. They all played such a significant part in keeping Matt alive through some of the darkest periods of his life.

Meanwhile, a battle of a different kind continued: the ongoing battle with the 12th/16th. We had sought some civilian help through our local doctor who had sent Matt to another specialist. The result was that both doctors saw the need for medication and further tests. We had even escalated the battle to Army HQ Tamworth by this stage. Matt thought it was worth a shot to see if the hierarchy above Armidale could offer him better medical support, order some tests, assess the new civilian doctor reports and not just take the scan done on day one as gospel, as Armidale had. But still no one would listen. Even when Matt went back, time after time after time, almost on his knees, begging for help, not a single person from the Green Machine offered any kind of support.

We had become more comfortable with letting certain people know that Matt was not well, people we felt would understand. You see, to look at Matt you would never know anything was wrong with him. In fact, he looked super fit and healthy, and unless you were on the receiving end of one of his outbursts, which no one ever was except me and, to a lesser extent, the kids, he came across as a lovely, happy-go-lucky guy. Only those who knew him really well and spent a lot of time with him would notice things, like his inability to assess time or his constant need to take off without warning.

Not only was I worn-out by this stage, but I was also losing faith that I had the strength to continue, especially not knowing what 'continuing' meant. It was all becoming too hard. The father of my little babies was becoming too hard for me to manage and he was changing from being my partner in life, my husband, my best friend and my safest place in the whole world, to becoming a stranger who I had to protect. The Army had become the enemy, and that was daunting because how the hell was I supposed to fight them? So many roadblocks, so many battles and no solutions in sight. To give you some insight into the incompetence and negligent attitude of the powers that be at Tamworth, here is a copy of an email from April 2011, notifying certain people of rank, whose names I have blacked out for legal reasons, that they could not locate any of Corporal Page's medical records as far back as February. But still, no one seemed to think there was a problem!

From: [redacted]LT 1
Sent: Wednesday, 13 April 2011 15:40
To: [redacted]MRS
Subject: FW: Members records [SEC=IN-CONFIDENCE]

From: [redacted]LT 1
Sent: Wed 13/04/2011 3:31 PM
To: [redacted]
Subject: FW: Members records [SEC=IN-CONFIDENCE]

Sir,

Thankyou for your assistance.

Letter dated 15 FEB added as attachment.

Regards

[redacted]

From: [redacted]LT 1
Sent: Wed 19/01/2011 11:29 AM
To: [redacted]MS
Cc: [redacted]WO1 2
Subject: Members records [SEC=IN-CONFIDENCE]

IN-CONFIDENCE

Ma'am,

Two of 12/16 Hunter River Lancers members were admitted to your facility in November 2010.

These members were : 8508034 CPL M Page [redacted]

I am having trouble accessing their inpatient information, having tried three times now.

12/16 HRL medics were given some paperwork by your facility before the unit left Puckapunyal but [redacted] all I have for CPL Page is a CT report.

When I rang in late NOV 2010 I was told the information would be accessed and sent to the 12/16 HRL RAP.

I rang last week and a search was conducted and all that was found was a CT report for CPL Page, of which I have a copy.

Today I rang and spoke with Graham who said there was all new staff, but he thought he knew where the paperwork may be.

These members are being followed up by specialists who have asked for copies of inpt notes and D/C summeries from their admission at your facility before any further care can be conducted. As you can understand this paperwork is important for medical ongoing care and for DVA submission.

Can you please assist.

Regards

1/07/2011

It should NEVER have been this hard. We couldn't even get medical records from one point of reference to another to get a clear picture of what was going on. I have folders and folders of inept and inadequate procedures that overall aided in the failure of the Army's duty of care immediately following Matt's accident. The 12th/16th rained hell on us for no other reason than they were inept. And no one of serious rank wanted to own it, let alone help us fix it.

As Jess had predicted, Ayleigh decided that greener pastures weren't so green after all and she came home after living with her dad for just three months. Despite telling her she would have to commit to a year before she left, I let her come home because I couldn't bear for her to be unhappy. And I missed her terribly. But now it was Jess's turn. She was hating on us one month and desperate to come home the next; all fairly run of the mill issues in families with teenagers. But add to the mix, a husband struggling to exist every day, the financial burden of a failing coffee shop, managing staff employed to help Matt, protecting the kids from Matt's situation, keeping Matt in check and doing my best to track his every move while raising two babies and working and travelling, and you realise there was absolutely nothing normal about our situation. Although buying into the coffee shop franchise gave Matt a sense of purpose and was therefore a positive step forwards, every time he had to set foot on the barracks, the experience was so painful that it was like he took two steps back again. He couldn't keep going back to the place that was potentially killing him; it had to stop.

By now Matt was in a dark place every other day. The civilian doctors he had started seeing had him on anti-depressants, which came with their own set of unpleasant side effects. It was a case of trial and error. The best the doctors could do was bounce the meds from one to another or adjust the dosage up or down in an attempt to find the right combo. Sometimes Matt's condition was so bad he wouldn't leave the bedroom. I have to admit it was almost respite for me on those occasions because he wasn't raging down on any of us. You can imagine the guilt I had for feeling relieved that he was hiding from the world and us because he couldn't cope. But that time gave me a little breathing space; it allowed me to re-set, build the kids back up and tell them it was going to be okay and that Daddy just needed to rest. I felt like I was constantly walking on eggshells, what with my teenagers hating me one moment and Matt raining down on me the next. So many times, I nearly broke. Only those closest to us would know some of the truth. They got a glimpse into how bad it could be and stuck by us anyway. Tash, Matty B, Kate and Meli, in particular, will forever be in my heart. They were the ones who picked me up on the days I wanted out and pushed me back in because they knew how much I loved him. They knew pushing me back in was the right thing to do and I am so grateful to them for that.

It wasn't all bad. Whenever Matt was feeling okay, we still laughed, he still played with the kids, and they loved him and he them. We had barbecues with friends and Matt would be social and friendly with everyone at those events. But unfortunately, the bad times outweighed the good. Even if an episode lasted just twenty minutes or so, the scar left behind was massive. I remember going to see Matt at the coffee shop one time so we could walk up to the bank together, just to get him out for some fresh air, as it hadn't been a great day for him. As we were walking up the street, he told me he felt 'fuzzy', just a bit off. Next minute, for no apparent reason, he just snapped. Some poor unsuspecting guy on the other side of the footpath may have looked at him a certain way, who knows what it was that set him off. In any case, Matt snapped, and it was on. He wanted to kill that stranger, and I swear he could have had I not intervened and jumped in between the two of them. I pushed on Matt's chest, begging him to look at me, telling him it was okay, that he was safe, that this man wasn't a threat. In the next breath I yelled to this complete stranger, urging him to go, to just run and let us be. The poor guy was understandably shell-shocked but thankfully he did as I asked. He had no idea what had just happened or why, which makes two of us!

Matt had decided months before that incident that, while his brain might be damaged and he had lost his career in the Army, there was nothing stopping him from keeping fit. So, he found solace in heavy lifting. It was his way to zone out. Trouble was, this also meant he was now very strong. In fact, he was all muscle and his adrenalin was pumped to the max. It was a scary combination because it meant I had no way of stopping him when he was in a rage, at least not physically. When he had been pumping weights, he looked like a male kangaroo with his chest puffed up ready for a fight. I laugh about it now, but it was no laughing matter back then. That's exactly what he looked like during this incident with the stranger in the street.

Anyway, back to the incident. Once the stranger had taken off, Matt suddenly relaxed, stepped away from my grasp, looked at me like I was some kind of freak and casually asked, 'What the hell are you doing?'. Before I could answer, he took my hand and continued walking up the street. And just like that it was over. I wanted to hit him and scream at him, but there was no point. He wouldn't understand what had just happened, and bringing his attention to it would have only made him spiral back down the black hole. That was my life, every second of every waking day, not knowing from one minute to the next if and when he would snap.

Sometimes Matt would do the most hilarious shit, things normal people would never do. The boys – a combination of the guys at the coffee shop and his Army mates – would encourage him, as they thought it was hilarious. But what they didn't realise was that Matt had now completely lost all sense of boundaries. Whatever this 'thing' was, it was stopping him from processing risk, time, fundamental social cues and courtesies. So, as hilarious as it may have seemed at the time, there was a downside to it. Every time he did something out of the ordinary, funny or not, it was up to me to watch him like a hawk and deal with the fallout. It still is. I have no idea why this happened and to this day, no doctor has ever been able to fully answer that question, except to say that the brain is such an unknown entity. So, this was life. Some days we had a laugh, most days we had hurdles and dark patches that I had to drag us all through. There was no balance, because there was no pattern. There was no defined trigger and there was nothing I could put my finger on that made him better or worse, except when the Army came up in conversation. That most definitely made it worse.

The senior ranking officers from the units at Armidale and Tamworth, the men charged with Matt's welfare, made him feel like none of it was real, like it was all in his head and he was nothing more than a whingeing pain in the arse. They made him feel like he didn't have the right to ask for help and that he should be strong enough to just bury it all and get on with it. And because of that attitude, Matt tried in the early days to do just that, but it only made everything worse, not just for him but for all of us. He would come home from appointments and just shrug his shoulders and say, 'I don't think they believe me, babe'. I was toxic. And I was scared, for him, for us, for the kids, and for me.

I had two choices: give up completely, which would have spelled the end of our marriage and the kids losing their dad, or dig deeper, go higher and fight harder. For the record, divorce was never an option. I honestly believe if I had walked away at that point, Matt would have killed himself. I decided it was time to take charge. I was going to go higher, and I was going to get help. There was no way that this was how the rest of our lives was going to look. I knew I couldn't keep stamping out fires and I sure as hell wasn't losing my best friend. In the end I had no choice. I had to fight. I penned a ministerial letter to the Minister of Defence, Science and Personnel and Veterans' Affairs and to the Chief of Army, and, as you can see, I didn't hold back. In fact, it would be fair to say I let rip.

MY MINISTERIAL, PENNED ON THE 29TH JUNE 2011:

From: Page, Karen <Karen.Page@chep.com>
To: Warren.Snowdon.MP@aph.gov.au <Warren.Snowdon.MP@aph.gov.au>
Cc: Hurley, David LTGEN
Sent: Wed Jun 29 14:05:33 2011
Subject: Please Help My Husband

To Mr Warren Snowden

My name is Karen Page and I am married to Corporal Matthew John Page (8508***034). I am writing to you to ask for help. I am tired of being the silent wife that sits back and watches my husband try and get help via all the 'right' channels only to be left out in the cold with no resolution.

Matthew has been an Army Reservist since joining in April 2006 and during this time he has deployed twice, once in 07/08 to Iraq and once to Afghanistan in 09/10. In his last deployment he sacrificed the birth of his first child, not meeting her until she was five months old. It almost cost him his family and marriage, but he stood strong and did what he believed was right.

On his return from Afghanistan, he was selected to apply for the SAS and he left for Perth for the trial period. As it was only 12 weeks after his return from his deployment, he decided it was too much too soon to be away from his family again and decided to return home with the option to re-apply the next year. The reason I am giving you this background is because I want you to understand he has never been just a part time soldier. His mental and physical aptitude for the uniform has always been 100%.

Prior to November 2010 Matthew was a very strong and focused individual, both mentally and physically. He had completed all levels of study required to start his own personal training and rehabilitation business. Prior to November 2010 his business was growing strong. His determination to build the business and support his family drove him to achieve a great clientele in a very short period of time. He also had in the back of his mind the option to re-apply for SAS in the future and continue his work with the 12th/16th Light Horse Reservist unit here in Armidale.

In November 2010 Matthew was on a training exercise. He was standing out of the manhole of a Bushmaster when it went into a ditch. He was hospitalised for 3 days with neck, head and back injuries. Since the accident he has not been the same. Over the last seven months I have watched my husband suffer daily with no help or support from anyone within ADF or DVA.

He has lost his business due to the mental and physical challenges he faces every day since the accident. He has lost the chance to go back and re-apply for SAS as he is no longer physically or mentally strong enough. Matthew is always tired; his physical strength is nowhere near the same levels he previously had, and his cognitive capacity is significantly reduced. This results in him getting frustrated often and unfortunately it is his family that bears the brunt of this.

I am not a doctor, nor do I know the right terminology of his condition or the correct channels, but I do know that for seven months he has asked and begged people within DVA and ADF to help him, he has been passed from one person to another with no outcome and this has added to his levels of frustration. He was removed from a course due to his 'medical condition' yet was still rated as Med Class 1. He is down in Singleton today seeing another Army doctor, so they can down class him medically, although I fear this is more to do with covering up the mistake of either taking him off course or not down grading him earlier, than actually helping him.

All of Matthew's medical reports are on his file, including one from a Non-ADF doctor stating that his condition is real and serious with the possibility of it being long term and he advises that the effects can lead to the possibility of depression. I am bewildered and angry as to why Matthew's condition is not being taken seriously.

There were other soldiers injured in the same training exercise and they have not only been retained by ADF financially but have received all required medical treatment including surgeries. Matthew has not been supported financially, or medically, or even morally by anyone senior to him. Matthew has been treated like a white elephant in the room and I am at a complete loss to understand why. My husband is not a white elephant and I'm tired of him being treated like one, he deserves better from all of you.

As a wife who has had the worry and stress of waiting for my husband to come back from deployment I am stunned and annoyed that his commitment to the ADF of this country has been disregarded so shamefully. My husband has been altered by the accident seven months ago and needs help.

I hope that you will take up his case and do the right thing. I look forward to hearing from you very soon.

Yours sincerely,

Karen Page

And just like that, bells began to ring and people started to take notice. How sad it is though that it took the desperate plea from the wife of a broken soldier for that to happen. Looking back now, I remember the feeling of sheer desperation from not knowing where to go, who to turn to or what to do. When I wrote that letter I had never before felt so helpless. I was losing the man I loved. I was terrified that he would fail to come home one day. He was so lost. And he had no idea how to ease the pain and stop the seizures. Not knowing the trigger, he couldn't even make one happen when a doctor was watching, making it next to impossible to diagnose. As a result, we didn't even know that they were seizures. He would just disappear all of a sudden, mid-sentence or mid-step. What if it happened while driving? I'm not even sure he believed me in the early days that they were even happening at all.

MATT'S PERSPECTIVE

Okay, so life in general is resistant to change. I think it has something to do with our evolution and natural selection. And the environment provides the stimulus. Human nature and the collective protective measures that have sustained humanity cause us to resist change. This resistance has even built itself into the way we act day to day, the way we socially interact and the way we build our political and fiscal systems. Even the way we approach STEM (Science, Technology, Engineering and Mathematics) is guided by this instinct.

What I have also discovered is that outright stubborn resistance is rarely the answer to this problem. You can only tilt at windmills 'Don Quixote' style for so long before it leaves you so fatigued and mentally diminished that you will either give up in frustration or become so jaded it infects every aspect of your life. These moments are important. You see, life doesn't care about your plans or how you thought your life was going to go. Life just goes on and time is on its side.

I was twenty-seven. I had one kid in nappies, one kid on the way, and a household to help support with a traumatic brain injury and a broken body. I had to start hustling. So that's what I did right up until the Army, in all their wisdom, decided to pull me from a critical promotion course due to my medical classification. I was outraged and, like a cornered animal, I was injured and needed help. The Army refused to help me. They refused to even change my medical classification because they didn't want to have to manage my convalescence. And the one fellow who was supposed to make sure this was all administratively sorted was the same guy who removed me from the course. This left me at an impasse Army-wise, with no control and no champion to help me through the process. At some stage, one of my colleagues mentioned that I should look into my insurance for my PT business and, sure enough, I was no longer covered.

So I went from having a pretty decent Defence career and a profitable training business earning enough money to support my family, to no longer being able to run this business and a headquarters that wouldn't employ me for anything because of my medical classification (that still hadn't been amended) and unwilling to downgrade me because of the residual responsibility this would place on them. I was out in the cold with no means to support my family. I was outraged! I felt betrayed and, despite trying to bury it, I was scared shitless. Everything Kaz and I had sacrificed and worked so hard for all these years and these insignificant little shitbags couldn't care less about any of it. I was in their office every day asking, begging, pleading for help, and I was always given the same mediocre response, 'I will have a look into it' or, 'I'll bring it up at the next HQ meeting'. This was always delivered with great sincerity, despite the fact that they clearly had no intentions to actually help me.

Months had passed since my accident and, psychologically, I knew I was in trouble. The sheer weight of the burdens I had convinced myself I had to shoulder, knowing that I was just one man taking on the ADF, had taken its toll on me and I was coming apart at the seams and fraying at the edges. I was also in constant pain, had memory issues, was having seizures (undiagnosed epilepsy from the traumatic brain injury), as well as the myriad other behavioural and social issues that come with a traumatic brain injury. I suffered massive mood swings, became hyper-aggressive, started becoming paranoid, developed social anxiety as well as general anxiety, and I wasn't treating myself well. I was sleeping maybe two to three hours a night and was starting to look for fights when I was out and about. I didn't even care if I won or lost; I just needed to do something that was not sitting in a holding pattern.

It was a lot to take on. I had never ever felt abandoned like this before, certainly not by an institution that prides itself on mateship and strong leadership. They had abandoned their obligation to me and that broke me. It wasn't the wars or the humanitarian situation overseas or the grind of being on operations or training courses that broke my mind. It was the pure, almost aggressive mismanagement of my situation coupled with the fact that everything I believed about my contemporaries and leadership in the Army was proven wrong. The most courageous war fighters I know couldn't foster the moral courage to help me and leaders I respected and would follow anywhere couldn't muster the fortitude to make a stand. I had taken heat from leaders countless times. I'd been reprimanded; I begged, borrowed or stole to look after those to whom I was responsible, and I'd have given everything to complete mission for my bosses. I'd kept our readiness on high alert whilst trying to keep the people I was responsible for healthy and motivated. And not 1% of it came back when it was time to return the favour.

I wrapped up my business, fired all my clients and crawled into my cave to try and shut out the world. This doesn't mean I completely withdrew from society. You see, without sounding too clichéd, when in pain or dealing with issues, men need space and time to clear their heads. So we withdraw into ourselves to try to process the issues. The problem with this tactic when you are married or have daughters, however, is that they like to talk things out, share and express emotions. Needless to say, such conflicting coping strategies and techniques will inevitably lead to a non-constructive place. It takes many years and a really good relationship to negotiate these waters. Luckily for me and Kaz, we have both up our sleeve.

The trouble is, when something builds and grows to this extent, there comes a time when it has to blow. This is what led Kaz to writing a ministerial letter. This seemingly insignificant step finally broke the stalemate. Within twenty-four hours of the ministerial going out, Kaz had received emails and phone calls from the minister's office and multiple generals. Being a good little NCO, all this attention, and the fact that my name had come across generals' desks, was really disconcerting. I'll give you a little insight into this: one of the many things the army beats into you is to be the quiet achiever; to be extraordinary and be the epitome of the professional soldier without drawing attention to your accomplishments, your work or your troubles. Now generals were discussing me and my situation. But the deadlock was over. It was unnerving, but it finally felt like I could rest for a second. No one was going to be able to stop this now; it was out of my hands. It was now in big enough hands to be able to motivate the unit and the DVA to finally start helping me. This is when we met Pete Goom.

It was time to look at the change as an opportunity, not something to be feared. Kaz biting the bullet and making contact with the minister's office was the catalyst that brought us out of the darkness and on the path towards brighter times. Now to deal with the fear of change and embrace it as the helpful influence it was intended to be. This was not without its trials and dark patches. But Kaz, true to her nature, propped me up and helped me find the assistance I needed but was too afraid to reach for.

So it took a ministerial for people to sit up and finally take notice, a full seven months after the accident. I couldn't help thinking at the time, God forbid if this gets media attention! I didn't think for a second that these high-ranking individuals were working with Matt's best interests in mind. I honestly believe they just wanted to see whether or not I had a story worth reading, whether this could hurt a career, a political agenda or even the entire ADF. At this point I didn't care, as long as Matt got the help he needed and deserved. As long as the wheels were turning and the right people were listening, we had a chance to get Matt the right help at last.

The following day we received a phone call from a Warrant Officer in the Australian Army. His name was Peter Goom and he was assigned as Matt's case manager. However, Pete quickly became much more than that. He became Matt's welfare officer and our guardian angel. Right from the start, he took the time to come and meet us, to really get to know us. Pete had been around the block a few times and he had worked with a lot of broken soldiers and their families; he knew fact from fiction, and he knew how to deal with me instantly.

He listened, he gave me time to recover when the tears and emotions got too much, remaining calm and considerate while giving me space. Once I had finished telling our story, he looked me in the eye and promised he would stick with us until we got help. I believed him.

By this time, we had sold our little slice of paradise on the outskirts of town and had built a home in Armidale. We also had the coffee shop and our investment property and I was still travelling the world for work. So, at least on the surface, we looked pretty good. But no one ever bothered to scratch beneath the surface; no one ever asked if we were okay, until Pete Goom came into our lives. It was now several months post-accident and we were still no further along medically. Matt was still popping Panadol and Neurofen like lollies and there was still no clarity around when and why the seizures would happen. They continued to happen a few times a day and all I could do was pray like hell that they didn't happen when he was behind the wheel, even though he only ever drove down the road. I couldn't stop him from driving because he would just get angry. He now knew he was having seizures but burying his head in the sand was easier than facing this reality and, with no-one to help me get him to see reason, what could I do? Pete was going to be my only help and I was going to hold on to him for dear life.

Pete was willing to look beyond the superficial and see us for who we were as people. I felt that even if we lost the battle, we weren't alone anymore. Very early in the piece, Pete realised it wasn't just about Matt. He understood that the impact on me and our kids was massive. He told us about his partner and the journey she had endured, which was so much tougher than ours. She had lost her husband, who also served in the defence force, leaving her with two children to support on her own. Pete could empathise with us. He had his own story as well, and, as a case manager, I'm sure he could write his own book about the many broken soldiers and families he has worked with and assisted over the years. These men and women have helped shape our history. I'm sure we could all learn a lot from their stories. He gave me his phone number and invited me to call him anytime. I have no idea if he realised just how much that meant to me or how much I would take him up on his offer.

For the next twelve months, I called Pete too many times to count, sometimes just to vent, sometimes to cry. At times I begged him to make it all go away. I just wanted my husband back. I would cry like a baby, acting like a spoilt brat who just wanted to have their own way. Each time the floodgates opened, I would get so angry at myself because it felt like they were winning and breaking me in the process. I'll let you in on a secret. I HATE crying. But even more than that, I hate being defeated. And I hated losing my husband piece by piece. Pete copped it all. I knew he couldn't fix Matt, and he knew it too. But he gave me hope. Just knowing he was listening and that he had the ear of the people who could make a difference really helped, especially on the toughest days.

Matt now had major mental health issues, not because of what he had endured on the frontline, but because of the post-war process, the total disregard and lack of support from those who were meant to have his back. It is so important to understand the difference and that neither is more or less real or significant than the other. Men and women who have come home from the atrocities of war are scarred for life and the impact of those scars can cause horrific Post Traumatic Stress Disorder (PTSD). We don't do enough for our returned veterans, we never have. Matt was deeply traumatised, and not because of his time at war in either Iraq or Afghanistan. Matt was traumatised because he kept having blackouts, he was in constant pain from head to toe, he was losing time, and even his cognitive abilities and speech had started to become affected. And no one in the Australian Defence Force would help him and he had no idea why, because no one would give him a reason. No one seemed to give a damn. From the tone of their emails to their body language, they gave Matt the impression that they found him problematic and a nuisance because he was causing them too much work. Not only was he depressed and traumatised at this point, but, in my humble opinion, he was also suicidal at times. And no one would help. The Department of Veterans' Affairs (DVA) hadn't even acknowledged anything was wrong!

CHAPTER 16
The black demon

I call it falling into a big black hole, Matt calls it the black demon taking hold. No matter what name anyone gives it, ultimately it is the same; any mental health issue triggers an unyielding pressure on the sufferer, it can cause deep depression, and leave the sufferer second-guessing every thought and wondering if their existence is a hindrance to those around them. I could always tell when Matt was sliding, I still can, and ten years on I still don't understand all the triggers; I doubt I ever will. But at least I now know the signs to look out for.

In those early days I could see him sliding, but I had no idea how to stop him from toppling headfirst into that black hole; to stop the black demon from taking hold. We would all just have to hold on tight and hope we got through the next round. But I could tell that having the black demon constantly hovering over his shoulder, just waiting to consume him, was beginning to weigh heavily on Matt's mind. It's hard to explain what it's like when someone you love is losing their battle with depression. Only those who have been there, or are fighting this battle right now, would understand.

It was always so hard to know the right thing to say to him. What do you say to someone on the edge? 'Hang in there, babe, it will get better'? In the end I had nothing, except, 'I love you and I promise you that, no matter what, we are going to face this together and I'm going to be there with you every step of the way. I'm sorry this is hurting you so badly, but please don't give up on us, on you or on me'. It's all I had. I doubt someone like Matt would ever open up about it. I honestly believe that if the black demon had a strong enough grip on him, he wouldn't share his thoughts, he wouldn't ask for help, he would just end it. Unfortunately, this knowledge brought its own tortures for me. My fear made me desperate to know his every move, just in case. I figured if he got to the point of no return, then I was damn well going to be there to fight and bring him back.

There were also days when the physical and mental anguish became so unbearable that, like a wounded animal with nowhere to run, he would strike out at the closest thing to him, and usually that would be me. He never physically laid a hand on me, but sometimes words and actions can hurt just as much, so much so that there were times when I just wanted to run and keep running and never look back. I even made plans in my head to pack my bags, pack up the kids, and go. I considered the money I had and how far it would get me. I allowed myself momentarily to fantasise about having some time out just to be a mum and enjoy the kids without the constant worries. But then reality would set back in.

This was one of those days. Matt was in a dark place as he had just received yet another setback from the so-called 'medical experts' within the ranks. They kept sending him from pillar to post, but no one was running tests, no one was admitting there was something wrong and no one was pushing for a follow-up. They were literally just ticking boxes, losing files and shuffling paper, all the while ignoring everything we had been told by doctors in the civilian world. Matt was beginning to verbalise his desire to give up. He complained that the pain in his head was never-ending and the line between physical pain and mental pain had become so blurred that he could no longer tell what was real and what wasn't. This was all happening because the Green Machine offered no pathway of support whatsoever. They made him feel like the Army had suddenly become the enemy rather than an ally.

The morning had been particularly brutal, with Matt striking out in anger, sadness and confusion at everyone due to the pain, while I tried to explain to the kids that they needed to give daddy some space, all while trying to get them organised and out the door for the day. Because of the state he was in, I knew there was no way he could go to work and I would have to cover for him. This meant working a shift in a world I knew nothing about in a role I wasn't very good at. Every time I had to fill in for Matt at the coffee shop I was uncomfortable as hell the entire time I was there. I drove into the carpark, turned off the engine and lost it. I was broken, I had nothing left. Keeping strong in front of Matt took everything I had. This man I loved was losing the battle and it was becoming obvious that he was shutting down, and I had no way of knowing what to do or how to cope, let alone how to pull him through as well. Do you have any idea how many divorces happen simply because the family doesn't know how to deal with a loved one suffering from a mental illness? It's not only the sufferer who needs support, it's the people standing next to the sufferer. They are suffering their own battle, watching the nightmare unfold, not knowing how to help, what to do or where to go for support. Often, opting out is the only solution they can see.

I called Pete, for no other reason than I had no one else to call because no-one else got it like he did. I screamed, I cried, I begged him to make it stop or at least help me fix it. I told him that our own Army was now our worst enemy because we were losing this battle because of them. In my emotional state, I asked him why this was happening over and over again, even though I knew he didn't have the answers. I also knew Pete couldn't fix it. He couldn't make the Army accountable. But he listened and stood with us. Every time we tried to get answers, Pete backed us and pushed for them to do the right thing. He once told me, if I was waiting for the Army to apologise, then I should give up because it would never happen. However, he also told me that they would compensate, financially and medically, and sometimes that just had to be enough. If sharing our story could achieve just one thing it would be to have the ADF and DVA realise that the support network and processes they have in place for the mentally broken needs to be as important and as impressive and functional as the support they have in place for the physically broken.

MATT'S PERSPECTIVE

I cannot ever truly thank Pete enough for what he did for us, but I'm most thankful for the person he was for Kaz. He helped her when I could not, and I reckon I will be repaying the debt for two more lifetimes yet.

The following email provides further evidence of the disgraceful approach senior ranking members of the ADF took towards Matt's case. The writer almost offers condolences to the senior members of the Tamworth unit stating, 'I appreciate the effect this would have had on the unit'. God help me right now because all I see is blood! What about the effects it was having on us, on me, on my precious little babies who didn't ask for any of this, on my teenagers who just wanted Mum to stop crying and Dad to stop losing his shit? What about the effects on us? We never received an email saying that anyone appreciated the effects this must have had on us. They bounced Matt around so badly, and they couldn't even read their own path. Oh, and just for the record, we bought a coffee shop, not a PT business! He lost his PT business because your staff were incompetent, and he got sick.

From: [redacted]OL
Sent: Friday, 1 July 2011 14:28
To: [redacted]MAJ 1
Cc: [redacted]MAJ; [redacted]COL 1; [redacted]LTCOL
Subject: FW: [redacted][SEC=IN-CONFIDENCE:STAFF]
Importance: High

IN-CONFIDENCE:STAFF

[redacted] thank for the quick turn around and I appreciate the effect this would have had on the unit.
Couple of RFI's that flow from the QA:

- As of 24 May he was recorded as MEC 1 and remains so on PMKeys - has this changed recently? (para 24 refers)
- When CPL Page was reclassified unfit on 8 Dec and subsequently paid ARTS what was his civilian employment at the time and was he subject to ESPS (re para 6 to 8)?
- QA indicates he was again assessed as fully fit on 18 Jan (para 9) - was he supported on ARTS from 8 Dec through to 18 Jan inclusive on ARTS?
- Pls provide a timeline of when he was spted by ARTS due to his injury and when he wasn't (due to being deemed 'fit for duty') - this quantifies the level of salary spt provided by the unit and when
- Pls confirm who determined a medical classification review was required (by a MO, Singleton Medical Centre?) and when it was done (if it has been done yet)?
- Pls confirm member bought a 'Personal Trainer Business' in March 2011 (not 2010 - seems unusual to buy that sought of business if your suffering the effects on injury)
- Pls confirm whether CPL Page has submitted any DVA claims and when they were submitted

Appreciate it is Friday afternoon so please have the above RFIs confirmed by 1400 hrs Monday to myself, cc LTCOL [redacted] above.

Thanks

[redacted]
LTCOL
SO1 OPS SPT, HQ 2 DIV

02 9349 0491
[redacted]

Intranet: HQ 2 DIV
Internet: HQ 2 Div
SharePoint: HQ 2 Div

Leadership – the art of turning mediocrity into excellence through personal example!

IMPORTANT: This email remains the property of the Department of Defence and is subject to the jurisdiction of section 70 of the Crimes Act 1914. If you have received this email in error, you are requested to contact the sender and delete the email.

From: [redacted] LTCOL
Sent: Friday, 1 July 2011 13:12
To: [redacted]LTCOL

7/08/2012

I remember a time when one of Matt's bosses from the 12th/16th had the audacity to come into our shop for a coffee. It was all smiles and handshakes like they were best mates. It made my blood boil while Matt just smiled respectfully. The Army indoctrinates them so much that, despite everything, he still had too much respect for the uniform and everything it was supposed to stand for to be any other way. It was only after his former boss left that Matt would crumble. And guess who was left to pick up the pieces and try to bring him back around again? It's amazing, how ten years on, the thought of that incident still makes my blood boil.

In July 2011, eight months post the accident and exactly four weeks after I sent the ministerial, the Tamworth 12th/16th – largely thanks to Pete – called for an IWB (Individual Welfare Board). It was to be a meeting between all the 12th/16th powers that be in Tamworth, Matt, Peter Goom and myself. That IWB meeting was to be the first of many and it proved to be by far the most significant because it was the first – and unfortunately the last – time I ever heard any person of rank own it, apologise and admit that they had dropped the ball. I remember every word the Major at Tamworth said that day. In fact, I will never forget them. His words were, 'We dropped the ball. Matthew fell off the radar because he wasn't downgraded early enough. The buck stops with me'. A formal agreement was put in place during that first meeting to improve the support process from that day forward. I felt like crying with relief. Unfortunately, that relief was to be very short-lived.

If only the Major was as good as his word. If only he had the decency and fortitude to actually do something more than just make empty promises. If he truly believed the Army was at fault and had dropped the ball, he would have made Matt a priority then and there. Instead, it would be another sixteen months of fighting before Matt was officially diagnosed. Sixteen more months of fighting every step, begging for the unit to make it just that little bit easier for him, for us. I was disgusted at the response from these people and I was gutted knowing every brick wall they put up was another kick in the guts to my husband and more demons for him to deal with.

Our story takes an interesting twist at this point. The Army decided, in their infinite wisdom, to send Matt to a specialist who seemed intent to confirm that if Matt had PTSD, it was a direct result of his time on the front line and nothing to do with the way he was treated and mismanaged following his accident. While I don't have hard proof that this was their plan, all I can go on is the hoops they made us jump through, and the questions this so-called 'PTSD expert' asked when we went there. But ultimately the Army was proven wrong, again. It was like a dark comedy act, only we were the punchline and there was nothing funny about it because my husband's life was on the line.

CHAPTER 17
Giving up

By November 2011, a full year after the accident and four months post our first IWB, Matt and I found ourselves having to jump through some pretty suspicious hoops. We were being bounced from Coffs Harbour to Sydney, then to the Gold Coast. Each specialist had a different theory, and each would write their reports, suggest different medications, and then question why Matt was prescribed the medication he was on. It was relentless. It seemed like he was being treated as an isolated case study by each specialist when they should have been working together as a team to help him. It honestly felt like we were chasing an enigma with Matt dealt another blow every time his medication combo was changed, as it would signal the start of yet another battle to allow the meds to settle with his other meds and be alert for new side effects.

Through all of this, and for each trip we had to make, the Army seemed determined to make the process as drawn out and difficult as possible. Each time we were sent to a remote location, it would challenge our time, money and sanity, all of which we had very little of at this point. We would have to organise someone –usually Grandma – to move in and look after the kids, as this was not a process any of them should have had to be confronted with. We also had to put processes in place for the coffee shop. Thankfully, we knew we could trust our senior staff to have our backs. Pity we couldn't say the same thing about the Army! We had to rely on Pete to work magic to get the 12th/16th to assist with Matt's transport, accommodation, meals, and any other basic needs. Everything was a constant fight with this unit. It was like they didn't *want* him to get better. Either that or they didn't believe anything was seriously wrong. Or perhaps they just didn't care one way or another. I have never been able to figure out what their motives were. It felt like they adopted the approach that if they just ignored the situation, it (and Matt) would just go away.

My husband was deteriorating before my very eyes. Every time the ADF made us jump through another 'specialist' hoop, they would make it hard for us to get there. We were constantly met by a series of brick walls whenever we tried to process requests for travel assistance. They made us fight every step of the way. I don't know what it was that kept Matt going through all of this. I can only assume that it was a combination of family and his own inner strength. Whatever it was, I will forever be grateful because, as we have all seen and heard in the news, most soldiers who have taken their own lives have done so because the war at home seemed an impossible war to win and a harder fight than the one they endured on the front line.

MATT'S PERSPECTIVE

This process dragged on and it got to the point where it was having an effect on my sanity. I started to believe that I was causing some people grief, and that something might happen to me or that they were going to find something that would not only be against military law but also illegal at the civilian level. It honestly felt like there was a conspiracy to shut me down, whatever it took. I even went so far as to put measures in place to 'punish' people who I was certain were responsible. Luckily this is all behind me now.

The following email gives you some insight into the lack of care shown and taken by the 12th/16th unit. Notice how Pete simply requests in the email that the unit advise on arrangements made for travel, accommodation and incidentals for Matt? All the unit was worried about was who would be footing the bill. Well, at the end of the day fellas, I'm pretty sure that, as a tax paying Australian, it was me!

To: [redacted] WO2
Cc: [redacted] MAJ
Subject: RE: Matt Page [SEC=UNCLASSIFIED]

UNCLASSIFIED

Greg,

I sent you this info on 2 Nov 11:

The bills should be sent from the treating doctor to the members Defence Medical Facility (your RAP or doctor the unit uses), he can then pay the bill or forward onto Joint Health Command for payment.

So I don't know why the bills have been sent to DVA, did they indicate that they were able to reimburse?

Regards

Peter Goom

PETER GOOM
Warrant Officer Class One
Army Casualty Assisting Officer
NSW and ACT
HQ 5 BDE
HOLSWORTHY Barracks
NSW 2173

Tel: 02 8782 2888
Mobile: [redacted]
Fax: 02 8782 3031
email: peter.goom@defence.gov.au

Have you visited the Wounded Digger Web Site yet?
http://intranet.defence.gov.au/armyweb/sites/woundeddigger/comweb.asp?page=201116

IMPORTANT: This email remains the property of the Department of Defence and is subject to the jurisdiction of section 70 of the Crimes Act 1914. If you have received this email in error, you are requested to contact the sender and delete the email.

From: [redacted] WO2
Sent: Friday, 18 November 2011 14:45
To: Goom, Peter WO1
Cc: [redacted] MAJ; [redacted] LTCOL
Subject: RE: Matt Page [SEC=UNCLASSIFIED]

UNCLASSIFIED

Sir,
I have spoken with the OPSO and CO and some questions/queries have arisen.

1. What Financial Authority do we have to be expending the funds?

2. The units budget is limited as it is, we are unable to support the travel.

3. Who is supposed to be responsible for all these appts and reimbursing of the money?

7/08/2012

Regards,

Greg Phelps

G.L. PHELPS
Warrant Officer Class 2
Chief Clerk
12/16 Hunter River Lancers

: (02) 6766 2367
: (02) 6766 2209
: [redacted]
mailto:Greg.Phelps@defence.gov.au

IMPORTANT: This email remains the property of the Department of Defence and is subject to the jurisdiction of section 70 of the Crimes Act 1914. If you have received this email in error, you are requested to contact the sender and delete the email.

From: Goom, Peter WO1
Sent: Thursday, 17 November 2011 17:10
To: [redacted] WO2
Cc: [redacted] MAJ
Subject: Matt Page [SEC=UNCLASSIFIED]

UNCLASSIFIED

Good afternoon Greg,

I rang Matt [redacted] today as a follow up to see how things are tracking and as a result there are a few issues I would like to discuss with you.

At this stage I understand the money that Matt [redacted] paid out for specialist appointments is yet to be reimbursed. Can you advise where the unit is with that reimbursement.

[redacted] Matt [redacted] travelling to Coffs Harbour on 23 Nov 11 for an MRI. Can you advise on the travel arrangements, accommodation and incidentals for that trip.

There is further testing to be conducted on the Cold Coast which will require the above arrangements as well.

Thanks for your support in this matter.

Regards

Peter Goom

PETER GOOM
Warrant Officer Class One
Army Casualty Assisting Officer
NSW and ACT
HQ 5 BDE
HOLSWORTHY Barracks
NSW 2173

7/08/2012

To respond with, 'This unit's budget is limited as it is, we are unable to fund this travel' was utterly ludicrous. For starters, Pete had given them specific instructions on how to charge the costs sixteen days prior to writing this email. And secondly, how is it possible that the unit responsible for the welfare of their own couldn't fund the costs associated with medical attention? By this time, I was at a complete loss as to how a person held in such high regard by two previous Warrant Officers, could now be treated with such disdain by the WO2 currently in charge. When the WO2 wrote – and I quote – 'We are unable to support this travel', he had copied in the Major in Tamworth. Remember him? He was the one who said the buck stopped with him. What a crock of shit! None of them ever had Matt's best interests in mind.

Well done 12th/16th for showing your inept ability to look after your own over and over again! You all made Matthew's condition far worse than it ever needed to be. You played a significant part in the demise of my husband, failing in your duty of care for no other reason than you were either too lazy, too disinterested or too reprehensible to care. I honestly feel sorry for anyone reporting to you or serving with you now. You are all collectively such a small part of the Australian Defence Force, and you are insignificant to the majority of people in this country, yet you collectively very nearly destroyed my husband, and none of you ever owned it. To me, to my husband and to our children, you will forever be the men who nearly destroyed one of their own.

Every single time they fobbed us off, Matt fell harder, his self-worth drawn into question. Each hit dealt a blow to our family as well. I felt like a broken record, 'Daddy isn't well and just needs to be left alone in the bedroom for a while,' 'No Willow, I'm so sorry baby, but Daddy can't take you to kindy today and he can't play with you', 'Hi Tash, it's Kaz, Matt's in a black hole again, so he'll probably hide out the back all day. Just call me if he goes walkabout please'. This was my life on repeat, protecting Matt from the world, protecting the world from Matt and somehow in the middle of it all, trying to keep my job going so we could eat.

There were some days the fight almost didn't seem worth it. On those days, Matt would say, 'Let's just stop, babe, I'm busted. They don't think I'm worth it and I can't fight them on my own. I gave them everything, I don't have anything left to give'... For fuck's sake, how was I supposed to respond to that? Why don't they teach us what to say when they break our husbands to the point of no return? So, I just said NO. I said the only thing I knew to be true and right. I told Matt that we were not giving up. I told him we would force them to help us fix him. I told him to hang in there with me, for us and for our kids because we all needed him. What else could I say? And then I would panic. I couldn't make him stay home; I couldn't tell him that wherever he was going I was coming too. He would often disappear for hours at a time and I would hold my breath and wait, either for him to come home or to get a call or a knock on the door to say he wasn't. Sometimes Matt and I would end up fighting because I would dig my heels in and push his buttons. Other times, his anger was beyond his control and he was like an injured animal with nowhere to run. At those times I would try to shield it from the little ones. But I knew they could feel the tension, hear the yelling and see the tears I couldn't hide. Jess and Ayleigh had to deal with this on top of their normal teenage issues. I tried my best to explain what was going on, but I'm sure I screwed it up at times. There's no rule book for any of this. When you are thrown in the deep end, you just have to sink or swim. Some days we sunk, others we managed to tread water.

The daily battle Matt endured was horrific, but the battle was just as hard for me, the person supporting the broken. I barely survived each day. I got through, but only just. This story would have been very different had I not had the support of some very special friends, people who didn't judge and just sat quietly with a listening ear and a shoulder to cry on when I needed them. And not one of them ever judged Matt either, despite the way he treated me at times. No matter how hurtful the shit he said to me could be, the thought of not having him next to me, the thought of getting through everything we did to build our family only to lose it all now, was not an option. Granted there were days when I just wanted to run, when I wanted the pain to stop, when I wanted someone to just swoop in and make everything okay. I just went through the motions and hoped to God he hung in there long enough for us to find the right help and for the Army and DVA to start giving a damn.

MATT'S PERSPECTIVE

My good mate Al Bungate, a man of the highest professional standing, said it best. There are two types of courage: moral and physical, with physical being the easiest. Unfortunately for me, these members from the 12th/16th unit were the epitome of cowardice, forsaking the needs of their soldiers to cover their own arses.

I knew we had two choices: give up and go our separate ways or push through harder than ever. We had to learn who this new Matt was, and I effectively had to change who I was. I had to learn to not take it so personally; I had to learn when to switch off and when to react; I had to learn how to communicate with all the kids about how to deal with our situation and what to hide from them, knowing I wouldn't always get this right. This single armoured vehicle accident shattered Matt's life and in turn shattered mine and had a ripple effect on all our children. We all had to adapt. But we chose to do it because the alternative of giving up, of not being together to watch our family grow, and for me, of not being with the person that could still make me laugh over nothing at all, was too much to bear. The thought of losing all that was a far greater fear than anything we risked losing through the fight. I would not let them, or this thing that was now all-consuming in our world, destroy us. I just didn't know how at this point.

CHAPTER 18
My darkest fear

December 2011. Reflecting back three days out from Christmas, I realised the past few weeks had been particularly tough on everyone. There wasn't a lot of joy in our home. While the tree was up and we were trying to keep it as real as we possibly could for the babies, the two older girls knew it wasn't a happy time and they were feeling the pressure as well. We were all just trying our best to put one foot in front of the other and keep our heads above water.

Matt had been through yet another round of rejections from the 12th/16th, making it difficult for him to seek medical attention. Pete was doing all he could to mitigate the situation for us, but even he wasn't getting through to the unit. It seemed nothing was going in Matt's favour and the stress and uncertainty of it all coupled with the constant thoughts in his head that this was what the rest of his life would look like weighed more heavily than normal on his mind. He was falling deeper and deeper into depression, the ever-present dark cloud hovering over him growing more ominous by the day.

Matt was no longer working out, which was unusual for him as it had been his solace before. And he was avoiding the coffee shop floor by hiding in the office out the back. When he was home, he just sat on the couch, avoiding engagement with any of us. Even conversation with me was kept to a bare minimum. Like a gauge, his mood determined whether or not we would try to strike up a conversation. If he was angry, we all tried hard not to ask him too much or get him to talk because it didn't end well for any of us. If he seemed sad, we would try and get him to open up a little or encourage him to play with the little ones or just get them to sit and cuddle with him. Where this would have had an uplifting effect on him before, he would now move away or disengage. I had no idea what to do and no clue where to go for help.

The coffee shop was also causing significant financial strain on us and I was doing everything I could to juggle the finances every week, robbing from Peter to pay Paul. It was a never-ending cycle of stress. I'm lousy at pulling a poker face, so Matt could see the pressure I was under and that made matters even worse for him. I would do everything I could to try and stay positive, but it was all getting too much and, by this point, we were both breaking under the stress of it all. While he had been given medication to help with the seizures and other medication to help with the pain, the combination created a dangerous cocktail that messed with his mind even more. Worst of all, he still had no real diagnosis. Every doctor he was sent to seemed to be collectively flying blind and going around in circles. The stress of not knowing combined with the financial stress was taking its toll.

To add insult to injury, the Army decided, in all their wisdom, to demand that Matt parade. This required him to don the uniform that reminded him of the hopes and dreams he had lost, return to the place that caused him the greatest grief and face the people he believed were trying to destroy him. I remember taking a photo of him in uniform. It was around the anniversary of his homecoming from Afghanistan. I wanted a then and now photo of him holding Willow. The photo showed a broken man, his eyes sunken and dark, his weight low, his posture weak. When I look at that photo today it reminds me of what I came so close to losing.

By now, Matt had begun to lose sight of reality, the lines blurred between what was real and what wasn't. The voices inside his head were a particularly nasty side effect of the cocktail of medications he was on. He described it as having five people talking to him at once all day long. It was relentless and it was driving him insane. He would tell me in despair, 'I just can't make the noise stop'. I felt completely helpless. I had no idea what to do. And I was angry. It was beyond me how any doctor could have prescribed anything at all when they still didn't know what the hell was wrong with my husband!

December 22nd, 2011. I was at home with the babies. Matt had had a meltdown, stormed out and taken off in the car. He had been gone a while. I called the shop to learn he hadn't showed up for work. He was supposed to be on shift and, as always, our senior staff had stepped up and managed things for us. It wasn't the first time this had happened. But it always created fear and panic in my mind. I knew I had to try and stay calm, maintaining as much normality as I could for the kids. I called a couple of people we knew, but no one had seen him. I held my breath and waited.

The phone rang and Matt's name popped up on the screen. 'Thank God,' I thought to myself. I answered the call but all I could hear was silence. I kept repeating, 'hello' and, 'Matt, are you there?' but still nothing, my nerves mounting with each passing second. Something wasn't right. Fear suddenly gripped me like a vice and I felt frozen on the spot. It took everything I had to keep talking, trying to sound calm and composed. 'Babe, I know you're there and it's okay. Just say something, please just talk to me, babe please. It's okay, I promise. Please just talk.'

Finally, I heard a broken voice sob through tears, 'I'm sorry, I'm so sorry. I can't do this to you anymore. You will all be so much better off without me.'

I knew it was Matt, but this was not my husband. This was a scared, angry and confused soul, sounding beaten to a pulp, utterly broken and lost. All I could think of in that moment was, 'No, no, we won't be better off. Please don't do this to us, to your kids, to me, to yourself. Please don't'. At the same time, my head was screaming. I didn't know whether I wanted to kill him or hug him. I panicked, and I got angry, so angry. I took a deep breath, held my anger in check and said to him, 'Please don't do anything stupid. Please don't do something you can't come back from.'

Right then it hit me. There was no coming back from this. There's no coming back from death! That's when the anger took over. I yelled into the phone, 'If you kill yourself, your children will have to live with that. You will have done that to them. Do you understand that? And I will make sure they know it. I will make sure they know you took the easy way out instead of choosing them, instead of choosing to fight for them!'

I was met with silence, broken only by his heavy breathing. Finally, after what felt like an eternity, he almost whispered, 'I'm coming home' and hung up. I knew he was angry at my outburst but at least he said he was coming home.

I can't tell you exactly what he meant that day, not for sure, and we have never spoken of it again. But I can tell you that I have never heard that level of desperation or emptiness in anyone before, not ever.

I needed him home. I needed him to be there for our kids and for me. I'd take him any way he came, as long as he was breathing. I knew we would get through anything, except death. Death is final. Suddenly adrenalin kicked in as panicked thoughts began to swirl though my mind. 'Where is he? How do I find him? Do I call him back and keep him on the phone while he drives home? Is he really driving home?' My head was in chaos. I felt a combination of panic and anger. I just kept thinking over and over again, 'You can't do this to your kids. You can't leave them to grow up thinking they were not enough for you'. I paced the floor, I cried, and I paced some more. I did my best to console Willow and Scarlet, who had witnessed the entire scene unfold because there hadn't been time to move away or check who could and couldn't hear me. Logic and reason had gone by the wayside. All I could do was react in the moment to what Matt was telling me. Two sets of innocent eyes looked up at me for reassurance. They were both too young to understand, yet they were old enough to feel all of it. I took another deep breath, reassured the girls and resumed my pacing.

It felt like several hours had passed, but it was probably only twenty minutes, when I finally saw his car head up the driveway. As soon as he set foot in the door, I flung myself at him and grabbed him tight. There was no anger left, only immense relief. He was here in my arms and, at least for now, we were going to be okay. I cried some more. I had no idea what to say. What do you say to someone in the aftermath of such a crisis? Thinking back on it now, I honestly don't think he would have called me if he was beyond the point of no return. I believe he would have just gone ahead and done it. But then again, how does anyone ever really know?

I asked Matt if he wanted to add his perspective to this chapter. He declined, not because he has anything to hide or because there is something about this part of our story that makes him feel ashamed. It's just that it was probably the lowest point of his life. His view on what happened is simple.

'I wanted to die because the pain was beyond anything I thought I could bear. There is nothing else here I can add to this. There are no other words that can describe it. I just wanted to die.' I can't describe in words how gut-wrenching it was to hear Matt say those words.

We all lost something during this part of our story, but no one more than Matt. However, we all gained something too; we gained a firsthand perspective of what the horrors of a war on home soil can look like at the hands of mental illness and what it takes to fight the enemy you can't see.

Three days later we celebrated Christmas. It was quiet and the mood was anything but joyous. But I have to say, I was deeply thankful. Every Christmas morning since, when we wake up and do presents and enjoy a special breakfast together, I quietly say thanks – to no one in particular – that all my girls have their dad by their side to watch them, to laugh with them and to help celebrate what should be one of the happiest days of the year. I say thanks that another year has passed, and we have survived the black demon again. Then I think about the ones who have fallen, and the ones they left behind. Every Christmas until the day I die, I will be both thankful and sad.

CHAPTER 19

Abuse is abuse!

In the new year of 2012 yet another IWB was called. This time I decided it was time my voice was heard too. I checked with Pete and asked if I could write a statement and have it added to the minutes. He cautioned me not to expect anything from it and I acknowledged his advice. However, I needed to be heard. As well as fast track a better outcome for Matt – for us – I also needed to start my own journey to recovery. In my mind, the only way to achieve both outcomes was by letting the people who had caused the greatest damage know what they had done.

I desperately hoped someone sitting in that room that day would take responsibility and help me push for change. But to own it is to be responsible for it and no one wanted to be held responsible. Pete was right. It did nothing for our cause because, to this day the 12th/16th have never taken ownership, never acknowledged that their inept attitude towards Matt's recovery and rehabilitation prolonged and worsened his condition. But at least I had it on record.

On the next page is a copy of the letter I wrote that was added to the minutes. They all sat in the room while I read it aloud. I was thanked for my honesty and assured things would get better, that improvements were being made to ensure better care for Matt and our injured soldiers. Trouble was, those people never came good with their promises. We never saw a single improvement let alone better treatment of Matt. And so the battle continued.

ANNEX A.

I wanted to have my say today; I wanted to give insight in a formal IWB on the impact to me. Im going to fight, im going to make what we lived through count. I want this on record.

On the 25th November 2007 my husband deployed to Iraq, I held my breath for 8 months hoping he was safe and would come home. In September 2009 when he deployed to Afghanistan, I again held my breath and waited for his safe return home. It wasn't my choice to live like that, I married a man who became a soldier and I supported the man.

He loved his work and sacrificed a lot because of his dedication and commitment to the job and I'm sure he isn't alone in that regard. I naively believed that should he ever need those that stood next to him and above him, that they would do the right thing and help, support and care for him.

The Chief of Army states on the Australian Army web site, the following:
One of my greatest responsibilities and commitments as the Chief of Army is to ensure soldiers who are seriously wounded, injured and ill while in service receive the best care and support to assist them in making a full recovery.

If that truly is his belief, why then did this not happen for Matthew?

On the 17th November 2010 everything changed and the commitment and dedication Mathew showed, the sacrifices he had already made, meant nothing, it was all for nothing and the words of the Chief of Army failed Matthew.

Through my career, when things have been tough, the corporate motto was "Take the emotion out of it and process it for the facts", and I have heard this several times from Army personnel over the course of the last 2 years.
Im still trying to find someone that can explain to me how?
How do you take emotion out of a nightmare situation?

How do I not cry when my husband couldn't remember his children's names,

How do I not cry and feel destitute when 3 days before Christmas (2011) my husband talks of his family being better off without him and im left wondering if he is coming home alive?

How do I not feel pain, when my husband is standing on a footpath wanting to physically beat the crap out of a passers-by and I'm standing in between him and a stranger begging the stranger to run whilst trying to get my husband to focus on me and try and calm him down, the whole time with him not understanding any of it and barley remembering it 4 hours later.

How do I not feel tired and worn out, raising two babies on my own, a teenager on my own, a full time job, a business and a husband who needed me around the clock just to function for the first 12 months post accident.

How do I not feel angry that through all of this and so much more, no one within the 12th/16th Hunter River Lancers or the Australian Army has ever stepped up and owned the nightmare in which they made us live through?

So here we are again today at yet another IWB, and whilst I have the utmost respect for both Jamie Harper and Pete Goome for again forcing you to face this situation, I am still to this day emotionally and morally disgusted with the Hunter River Lancers and the Australian Army as a whole as nothing has really changed no one has been held accountable and no acceptance or even formal acknowledgement that a mis-conduct ever took place.

There was considerable media attention around this time relating to an independent inquest into the mistreatment of soldiers based on allegations of sexual and other abuse in Defence. The inquest was called "DLA Piper". I did some research and submitted an application, but it was rejected as Matt's situation was not deemed abuse. I was devastated. I didn't understand how they felt this wasn't abuse. Matt's was a case of abuse of power and negligence. The ADF had clearly failed in their duty of care towards one of their own. How could they honestly tell me this wasn't abuse? Why did his situation not count as much as someone who had been sexually abused while serving in the Defence force? Why were Matt's mental and physical scars deemed insignificant? If "DLA Piper" was not the right channel, then why didn't they offer us an alternative? Why didn't they at least suggest other avenues of assistance? Why is there still no support for families of broken soldiers or the fallen by suicide? We hear and read about it in the news all the time: reports of family members left behind, struggling to come to terms with the death of their loved one and, when asked why they didn't seek help, the response is always the same: 'I tried but I was passed from pillar to post and no one actually helped me'. That was my story too. I had no clue where to go. I had Pete in my corner. However, he was but one man in many corners because he was also helping other families just like us. He could only do so much. I wrote the Ministerial and, although that got people talking, we were still hit with massive barriers and struggles beyond belief, to the point where nothing actually got done. Where was I supposed to go? Who was I supposed to turn to?

We were back to square one, once again being bounced around with doctors changing his meds every other appointment. And the powers that be continued to make it difficult for Matt to make his medical appointments by withholding the funding to support his travel and accommodation, all the while continuing to ignore the fact that they had destroyed his military career. I felt useless. I felt like I wanted to rip someone's head off and shove it up their arse most days. Other days I just felt like walking away from it all. I have a strong personality and a deep sense of what is right and what is wrong. And it annoys me on the best of days when incompetent, inept, useless or even malicious people are doing the wrong thing, and getting away with it, causing good people to suffer. But this was more than that because it was personal to the core, and there was absolutely nothing I could do to stop it. Knowing how much Matt was hurting, I lied to him. Every day I told him I was fine and that we would get through. The truth was, I didn't know how I would survive let alone how we would get through.

February 2012. Not long after the latest IWB, we received notification that we were off to see yet another doctor. By this stage, we had already travelled from Armidale to Tamworth to Singleton to Coffs Harbor and to the Gold Coast. Now they were sending us to Sydney to see a doctor who was apparently a leading Australian PTSD expert. So commenced the usual epic process of arranging everything back home so we could make the trip. If this guy was even half as good as his reputation, we might just have been on the right path. So we decided to do whatever it took to make the appointment. I'm forever grateful that we had Grandma, who was happy to make the four-hour journey to come stay with the kids while we were away. Cherie never once complained or said she was unable to help. Every time we asked, she knew it was important and would just pack a bag, book a train ticket and stay for as long as we needed her to be there. And the girls loved it when she stayed with them. They would cook with Grandma and have special snuggle time, read stories and do everything little kids love to do with grandparents. It was the only thing that made it okay to be away from them.

We booked a place to stay the night in the city and invited my parents to come and meet us for lunch before the appointment. It had been almost a year since they had seen me and, in that time, I had cut all my hair off and lost a lot of weight. Last time they saw me, my hair was dark brown and very long, literally down to my arse. Now I was rocking a platinum blonde crew cut and was fifteen kilos lighter. We found a car park and pulled in. As we were getting out of the car, I saw my parents walk past us. Dad looked at me and gave a half nod, as you do when you walk past any random stranger, politely smiled and kept walking. I laughed and asked if perhaps he would like to try that again. He stopped, turned back to look at me as if he was about to go off at this random stranger, but then he nearly fell over when he realised it was me. We both laughed and went on to have a really nice lunch. It's the last nice memory I have of my dad.

It was time for Matt's appointment with the new doctor. I remember sitting in the waiting room, talking to an older gentleman. He told me he was a Vietnam vet who was, like so many others, still trying to get acknowledgement, compensation and medical help for the horrors encountered during and after the war. I was shocked. I sat with this man for an hour, listening to him tell me about all the troubles he had trying to get help. My heart broke for him. I wondered how, as a nation, we could do this to him. How could we ask him for so much and then make it so hard for him later in life? Why did he have to fight so hard for my freedom and yet, sitting here with him now, there was nothing I could do to make it easier for him. All I could do was listen and let him feel like he had a voice, and someone gave a damn, even if only for a short while. His story made me wonder why we still can't make life easier for a couple of thousand soldiers, the ones who are left, still struggling from the atrocities that fighting in a war leaves them with. Our government at the time of Vietnam chose to send men and women across an ocean to face a war no one wanted and then brought them home to a nation that treated them like dogs.

How did we get it so wrong? How are we still getting it so wrong? Why do we spend so much time and money on making sure we have the best of the best killing machines, but then when our servicemen and women break, we throw them out like broken armchairs not worthy of being fixed? We should be spending as much time, effort and money on repairing and helping them. Maybe then we would learn what's breaking them so we can focus on strengthening that part of the system before we send them out as killing machines. And for all the people out there saying I shouldn't even suggest making them killing machines in the first place, I agree that in a perfect world, no country would ever go to war again and we would all live happily ever after and the bad people of this world would just disappear. Then I come back to reality and some idiot has blown up a hundred innocent people in the name of this or that. There are no winners here.

As a nation, we need a defence force to protect what many take for granted: our freedom and independence. There are many layers of complicated political red tape, ego and crap tied up with all of this and I am not arguing that one way is right and another way is wrong; it is what it is. So, we should do it right because these brave men and women deserve so much better. After all, they are risking their lives to protect us. Our government today is sorry for the Stolen Generations, and rightly so. Why then are our leaders also not sorry for the broken soldiers who fought for the freedom we enjoy? Why is our government not making it easier for these men and women to get help, live comfortably and be taken care of? To this day I don't understand how we got this so wrong. And we are still getting it wrong. We aren't treating today's broken soldier any better. Did we learn nothing from Vietnam?

Do you know how many soldiers have committed suicide since returning from war? Do you know how many are living on the street, are out of work, have lost their family, have lost their mind, or are existing in a broken body that no one wants to help? Irrelevant of which war he served in, irrelevant of how old the soldier, what religion he believed in, or what colour his skin, if he took his life, he did so because he felt he had nowhere else to turn. He gave up and we let him. How have we not changed? How have we not stopped this yet? When do we band together as a nation and say *enough*? When do we all demand change for them? I hear so many people tell me how sad it is when the media reports of a soldier ending their life and then ask why the government doesn't do more? We need to ask that question loudly and collectively and keep asking the question until they sit up and take notice. I feel my blood boil every time I write about this or have these conversations. But I would much rather the discomfort than the fear of not talking about it at all.

It was finally our turn to see the doctor. Matt and I decided long ago that I would be there by his side every step of the way. In any case, he was not strong enough right now to do it alone and I was not having anyone, not even a doctor, put words in my husband's mouth. I also wanted to ensure that Matt told the doctor the full story, not leaving any detail out. Furthermore, to move forward, we needed accurate answers, and to get accurate answers all the doctors involved needed both sides, Matt's and mine. I found the doctor to be quite dismissive, almost rude. He may have been the best in Australia but right from the start, it felt like the aim of the consultation was to prove that Matt's PTSD was the result of his time in Afghanistan, not the injury sustained in an armoured vehicle accident on Army time followed by the inept attitude of ranking staff at the 12th/16th.

It is common knowledge that you never ask a soldier if they have ever killed anyone and you don't ask them for details of what they did at war. And if you didn't know that, then you do now. So, besides the nightmares Matt would have every night for the first three or so months after coming home, which were more like re-enactments than nightmares, I really had no idea what he did over there, and I never asked. If conversations were ever raised about what the boys did, Matt would just play it down and say it was a holiday to a sandpit, or, 'We just worked out every day'. In other words, he would brush it off and deflect the conversation in another direction. It was interesting to listen to some of the other boys handle it in almost an identical manner. It's like they were conditioned never to tell the truth about the horrors of war, whatever those horrors were. However, this doctor seemed to have missed the memo because he went there, and not in a subtle way either. It seemed this guy was going to do whatever it took to make sure he was 100% certain either way. And if that meant pushing Matt to snapping point, then so be it.

Matt is a traditionalist at the core of his being. He demonstrated that when he asked me to marry him, asking my dad first. And protecting me is always at the forefront of his mind. This doctor was pushing that boundary because he was asking Matt to talk about things he was clearly not happy talking about, not because he had an issue with anything he had done, but because he didn't want to discuss this taboo subject in front of me. The doctor persistently asked the same question over and over again, each time phrasing the question slightly differently. Matt would simply play it down, as he always did, and try to change the course of the discussion. But the doctor wasn't letting up. With each question, I could see Matt's discomfort intensify. He was starting to sweat, and his breathing was becoming more rapid and shallow. Suddenly, without warning, he snapped. Slamming his clenched fists on the doctor's table, he stood up, leant over the desk and blurted out a statement that will forever be left in that room. My heart was racing. I wasn't sure if I needed to jump in between Matt and the doctor – as you know, I've been there before – or just wait and see what happened. I decided to wait, and to his credit, the doctor simply leant back in his chair, never once breaking a sweat or flinching – as I'm sure he had seen far worse – and casually continued asking Matt a few more questions. At the end of the session, the doctor thanked Matt for his honesty and told him a report would be sent to the unit in due course, leaving Matt none the wiser and still stressed from the session.

We headed off to the city for a night to relax and re-set. This particular night was one of the only times throughout this entire journey that Matt really opened up about the what ifs. 'What if I never get better? What if this is as good as it gets?' The what-ifs were followed by questions like, 'What would it do to our kids and our family? How would we survive? How would we pay bills and put food on the table?' They were all reasonable questions that genuinely needed to be asked and discussed. But at the same time, they were questions to which neither of us had any real answers. The best I could say at that point was simply, 'I don't know. We will just make it right somehow for the kids and, as long as we have each other, everything else will be okay'. There was no other answer to give. Neither of us knew how to stop Matt's daily pain, let alone what the future held. We just had to take each moment as it came, remember why we were still together and ensure we remained true to ourselves as best we could. Whatever the outcome and no matter what happened moving forward, nothing would break us.

The report came in from the doctor a few days later. It confirmed that, while he believed Matt had depression, anxiety, and adjustment disorder, he did not believe it was the result of his service on the front line. Rather, he believed it was a direct result of how Matt's case was handled after the accident. Finally, someone was getting it! Besides Pete Goom, this was the first person who could actually see past all the political bullshit and realise this was a man in serious trouble if something wasn't done. And his report made that very clear.

The accident occurred in November 2010. It wasn't until February 2012 that a proper investigation was requested, resulting in a report stating that Matt had suffered significant trauma during the accident and the treatment received – or lack of – post-accident had had a detrimental impact on his ability to recover. They made us endure fifteen months – around 450 days – of living hell! And it wasn't over yet, not by a long shot. But for the very first time, there was enough written evidence – evidence that couldn't be swept under a rug – to support a case for proper treatment and the correct assistance from the DVA. We knew we still had a fight on our hands, but at last we felt like there was a beacon of hope.

2012 was another difficult year for us, but at least we had finally received medical recognition for Matt and were on a path to help him. Whether or not it was the right path remained to be seen. The process meant hard work for Matt. He would often tell me that one of the hardest battles took place in his own head, fighting the demons that kept telling him he was being a hypochondriac. I would tell him it isn't the demons in your head making you feel like that; it's the demons of the 12th/16th. And then I would remind him of the downs, the black hole, the seizures, the fights he would have with me. I would reinforce the fact that it's all real, even though we can't see it on an X-ray. It's heartbreaking to watch the man you love question every aspect of his life, questioning whether or not he had the right to even breathe. And it isn't just the demon lurking in the corner that guys like Matt have to deal with. It's also the daily reminder that they are no longer the men they once were; they are no longer capable of being soldiers. This begs the question in their minds: 'So, who am I then?'

I felt like I was way out of my depth in an ocean I had never swum in before. Pete was great at trying to push the 12th/16th to make life a little easier for us with funding assistance for travel etc., but what about everything else? I had no one to talk to and no idea where to go for help. How was I supposed to manage Matt's dark periods, and what on earth was I meant to tell the kids when they saw things they shouldn't see? How was I meant to help the kids through this whole process? I'm sure there was support available somewhere; I just didn't know where to look. And the people in uniform closest to this situation weren't exactly forthcoming with any information for me. Still, we kept going, one foot in front of the other. What else could we do?

Financially things continued to fall apart, and we were beginning to wonder what would be left at the end of the battle. We had bought the coffee shop as a safe haven for Matt, and that is exactly what it was. But it came at a huge financial cost. The business was not making money and it was hurting us. Everything we had saved was being swallowed up to keep the coffee shop afloat. Matt and I had no idea how to build a solid retail business. But it served an important purpose. So, in order to keep our staff and maintain this safe haven for Matt, all we could do was keep rolling whatever safety net we had into it until it became unsustainable.

Throughout 2011 and 2012 we endured two full years of fighting a battle we should never have had to fight, nor one we wanted. Many people told us to walk away, that the fight was too big. But then what would have happened to Matt? Where would that have left our family? Did anyone ever stop to ask that question when they were being so righteous with their opinion? Why should the Green Machine be allowed to treat the mentally wounded like this? Just because my husband didn't get injured overseas, just because he wasn't killed, just because he didn't lose an arm or a leg, just because those incompetent people couldn't see a physical wound, none of that made it any less real. So why weren't they required to act as a matter of duty to look after one of their own? The mismanagement of my husband was finally proven to have caused damage beyond repair. As I said earlier, and I meant this with every ounce of my being: abuse is abuse! It was high time the powers that be at the 12th/16th acknowledged it. Little did I know then that this day would never come.

CHAPTER 20
Nothing left

It is such a lonely path to take when you feel like you have nowhere to turn for emotional, physical or financial support. It was like being trapped in a room, knowing there was a door somewhere and not being able to find it, only to eventually discover there really was no door at all. It was one disappointment after another, and now we were facing the prospect of having to miss payments for the business. We were rapidly running out of money and options. I knew it was only a matter of time before we would either have to close the business or sell it, if there was anything left to sell that is. But neither was an option right now as I still needed Matt to have something to hold onto, to give him purpose.

By this stage we had moved from our little farm – my slice of paradise – to a residential block in town, mostly because of the business, but also because looking after seven acres and owning a business in town was too hard combined with the fact that it had become increasingly obvious that Matt was unhappy on the farm. Various fixtures in the house were starting to break down and we didn't have the spare resources anywhere to keep up with the maintenance. And to be honest, I didn't have the energy left to fight to hold onto it. Right now, Matt was the priority. Making sure he was in a safe place far outweighed any house, land or lifestyle I had wanted. So, we rented out the farm, bought some land in town and built a house on it. We thought adding to our investment portfolio was the right thing to do and that we would manage financially. But then the business started draining us, at first little by little, growing bigger as the weeks rolled on. We had no idea how to run a small business and it was really starting to show.

On top of this, we also lost the long-term tenant in our unit in Queensland and the market had dropped through the floor; if we sold now it would be a fire sale. We were at risk of losing it all and I was working as hard as I could to try and make it all work. I needed some breathing space to figure it all out. I decided I had no choice but to reach out to my parents, but it took me a few weeks to gather the courage to make the call.

I had never asked my parents for a cent before, but my brother had – many times in fact – and Dad had always helped him. But asking for financial help was really hard for me, especially from my parents. I'm not really sure why. It isn't like I had asked before and been knocked back. My father was a very hard man and his expectations of me were always far higher than I could ever hope to meet. And Mum and I just weren't close. She seemed to disapprove of my choices in life and made no secret of the fact that she thought I failed at everything I did. I think I had just reached the point where I was so tired of being judged by them that it was easier to stay away. They really had no clue what we were going through – they never really asked – and I was never forthcoming with information about how hard it was. If they called in my weakest moments, I would avoid the call and ring back later, simply to avoid being judged and hearing the words, 'I told you so' from my dad. I had literally looked at all other possible avenues to try and get through this without going to my parents, but I realised I had nowhere else to turn. I was desperate and completely out of other options.

We had decided to put the farm on the market and I just needed some respite and breathing space until the proceeds of the sale came through. Once that happened, I knew I could manage again and would be able to pay my parents back immediately. I took a deep breath and made the call. Mum picked up. I explained the situation and she listened and actually seemed okay, almost empathetic, which took me by surprise. Mum told me she would talk to Dad and call me back. Sure enough, Dad called a few days later. A bit of general chit chat and then we were down to business. He told me Mum had spoken to him about our situation and that they had discussed the options. Before I could respond he said, 'Yes, we will lend you the money. But one day you are going to have to learn to stand on your own two feet'. And there it was, all my fears voiced in a few careless words. It was abundantly clear that my father had given zero thought about how hard it was for me to ask for help and also not a single thought about the hardships Matt, the kids and I must be facing for me to get to that point of sheer desperation. He just wanted to have the last word and fuck anyone else. In that split second I froze, while on the inside my blood boiled.

'I've never asked you for money before, not ever. Yet your son has, several times. So, just out of curiosity, did you respond that way to him every time you helped him?' I asked my father. I didn't wait for his response because I knew what it would have been. All I said before hanging up was, 'Well, you can take your money and shove it up your arse! I'll find another way'.

I was devastated and heartbroken. And yes, right at that moment I had cut off my nose to spite my face. But the pain of his words drowned out any common sense I should have had. My parents had no clue about the depths of despair we were going through. While I acknowledge that I wasn't overly forthcoming with the details of our problems, they were never kept completely in the dark. It's just that what we did tell them, they either didn't believe, didn't understand enough to realise the seriousness of it or they didn't really care. My father's generation is of the belief that you just suck it up and that, as a man, you only show three emotions: happiness, anger and indifference. Nothing else exists and you certainly don't talk about mental health because that is what beer is for. Have a beer and she'll be right.

Matt was standing in the lounge room of our new house as I responded to my father. Once I'd hung up the phone, he walked over to me in silence and held me in his arms. It was exactly what I needed of him at that moment. I cried for hours because I had no idea what to do. I had run out of options. I felt like I had let my family down and didn't know how to come back from that. By the time my tears had run dry, I was just numb.

MATT'S PERSPECTIVE

What you need to understand, if it hasn't already become apparent, is that when the wolves are circling, Kaz and I don't hide or run. We double down. In the words of King Henry from Shakespeare's *Henry V*, 'Once more unto the breach' always seems to be our cry. Whether we were looking for a new house for our growing family or trying to buy and then run a small business, the balance of the scales was, as always, tipped towards Kaz. She was working her guts out professionally, managing her pregnancy and a toddler, as well as two high maintenance teenage girls. Then she got lumped with a broken head case that used to be her husband. I had my head so far up my arse trying to gain any form of assistance from the department that I couldn't see how heavily the weight of it all was resting on her shoulders. I see it now with the peculiar clarity that time and change of perspective can achieve. Not only was she handling it all, but she was somehow managing it all too. Kaz found a solution for everything until she couldn't, as you've probably gathered by now, so I knew she would push through to find one again. But this particular ask was such a huge challenge for Kaz, and the absolute feeling of betrayal and disappointment was written all over her face. It wasn't about the pain of asking for a helping hand and being told no. It was the absolute inability for Brian to just be a dad and see how hard things were for his daughter and how much it meant to have to ask for help. As expected, he tore shreds from Kaz's self-worth, just stomping all over it with careless abandon. And to top it off, it was just a given that he would lend money to his other child year in, year out, albeit probably with as much verbal venom. But this was different. This was the only time Kaz had ever asked him for a loan. I curse his name regularly, even still. He left a legacy, he justified so much in life that he was building a nest egg for the kids to inherit, and yet, realistically, he left a far larger legacy in the form of emotional trauma. I could live with most things from that man, but this is where he crossed the line, so much so that he still lives on in the insecurities Kaz battles with every day.

The universe must have been on my side because a few days later, we received a call from the local real estate agency to let us know that the tenants in our property wanted to buy the place. This news meant that, at least for now, we were going to be okay. My mum called me a week or so later to check in and let me know that, despite my words with Dad, they were moving ahead with getting the loan sorted for us and we would have to sign an agreement that was being drawn up for us by their accountant. I took great delight in letting Mum know that I no longer needed their help because our property had sold, so she could take the agreement she had her accountant draw up and burn it. I would never ask them for another dollar. I'm pretty sure Mum passed on the message, which probably infuriated my father even more. Dad had never liked to lose, and life to him was pretty much a win or lose, black or white game. He was going to sit and wait, and then, as soon as he found a reason to call and lash out, I knew he would. Sure enough, it didn't take long for that to happen.

Dad called later that month. Jess, who had been giving us grief at home, had moved in with my parents for a while. My father, being the man he was, must have felt the need to weigh in and let me know how badly he thought I was doing as a parent. I guess he didn't like losing face about the loan, so this was his payback. Dad's tone made it clear that Jess's antics were affecting them too. She was eighteen and old enough to know better, but somehow her behaviour was all my fault as a mother for not raising her better, stricter, or whatever the hell he thought I had done wrong. In any event, he felt he needed to tell me how and where I was getting it wrong. He was straight into me, having a go, venting and raging. How dare he give me his pearls of wisdom as though he rocked at the parenting game. It was the last thing I needed: more pressure, more failure, more insight from a supposed expert. As if the last four years of battles and hurdles hadn't been tough enough!

By the time Dad had finished his rant at me, I wasn't angry; I was just so tired. I had no fight left for this man. I had listened, but I had nothing in response. I felt boxed in. Yes, Jess was being a pain in the arse and making some poor teenage decisions, we all knew that. And yes, maybe I should have gone down and given her a lecture; maybe I should have made some different decisions at certain times throughout her upbringing. Hell, maybe he was right, and I *was* a bad parent and had got it wrong her entire life! But right now, right at this very point in my life, right in that very moment, I was dying inside. I was his daughter and I needed help and I had no one. And I'd had enough. All I could do was shut it down. So, I took a deep breath and calmly said, 'Your opinion would actually mean something to me if you were a half-decent father. But as we both know, there is no chance of you ever winning a Father of the Year award. So, you can take your opinion and, just like your money, you can shove it up your arse'. Before he could respond, I hung up. I put down the phone and sat staring into the abyss.

For the first time since the accident, I felt like I had lost. There had been times when I had wanted to run away and make it all stop, but there was a fire in me that always brought me back. Not this time. This time I just wanted my parents to help me. I just wanted my dad to stop being angry at me and I just wanted to feel safe, to feel like I wasn't responsible for every damn thing in the world. In that moment, I honestly couldn't remember the last time I actually felt safe. I took all of that out on my dad. In that one response I let all of that anger and hurt out, and sadly – although I didn't know it at the time – it was the last time my dad and I would ever go head to head again. In fact, it was the last time we would ever really speak at all. I later learned that my father was very sick at the time of that call. He hadn't let on to anyone, not even Mum. But she had her suspicions. It took a few weeks, but finally she convinced him that he needed to go to the hospital. By then he was so bad she had to call an ambulance.

For five of the six weeks Dad was in hospital he was conscious and confident he was coming home. I should have gone to see him then. I called him once, but he wasn't really interested in chatting. I think he was still cranky with me and probably also pretty scared. By that time, his body was battling many issues. I learned that he had lymphoma which turned into leukemia almost overnight. He also had heart issues – he was already part of the 'zipper club' – and Type 2 Diabetes that hadn't been properly managed over a twenty-year period. It was all killing him, just a lot faster now than any of us realised. During that one call, I was lucky to get three words out of him before he handed the phone to Mum. He didn't even say goodbye.

I went to see him on the sixth week. By that stage he was in an induced coma and I had to be the strength of the family. I had to be there for Mum while she went through a bizarre grieving process, completely shut down as though he was already dead. I had to be the one to watch my older brother struggle to come to terms with what was happening and totally incapable of stepping up to take care of business as the man of the family. So, the decision to turn off Dad's life support was left to me. I didn't get to talk to him one last time. I didn't get to tell him that it was okay and that we were okay while he was conscious. And yet I had to be the one to step up, to give the doctors permission to stop life support, to tell my brother he had to stop fighting for something that was never going to happen. I was the one who stayed with my father until he took his last breath on the 24th of June 2012. But I didn't get to say goodbye or tell him I was sorry, sorry that he didn't get it, that he didn't get me. I was sorry that he couldn't see how much we were dealing with, how much pain I was already in. I was sorry that he didn't understand how serious our situation was. Most of all, I was sorry that he didn't know me well enough, as his daughter, to know I was breaking. But no matter how sorry I was about all that, I will never be sorry for telling him to shove his advice and money.

Be careful what you say to people you love; you just never know if it will be the last thing you say. And I can say from experience, living with that memory of the last real interaction between my dad and me, is not good. However, I stand by the reasons I said what I did. I just wish we could have had the kind of father/daughter relationship that allowed us to talk. But you don't always get what you wish for.

He had his flaws, we all do. And if I could change one thing, it would be that he saw my struggle, that he realised how bad things were and supported me. Looking back now, I realise that, as a stubborn man, it just wasn't his style. And I guess the harder he was, the stronger I became. Dad, I'm sorry you never understood me and I'm sorry I never really understood you either. I love you and miss you every day.

CHAPTER 21
Sacrifices

While everything was unfolding with my father, the business was at the brink of collapsing and Matt was getting worse by the day. His struggle with the demon lurking in the shadows was relentless. It was a daily battle to keep him from falling down the big black hole. My work with CHEP had been ticking away in the background, but the role was now demanding more and more of my attention and time, time I just didn't have to give. I was still trying to figure out how to deal with Matt's dark times, while protecting the little ones, managing two wayward teenagers, working hard to put food on the table and doing my utmost to keep a failing business from closing its doors because it was the one thing protecting my husband.

Of all the challenges, the toughest by far was managing the demon. It still is today. Whenever Matt goes into a black hole, it's not just Matt I have to monitor and try to work with, it's also the kids I have to protect and our network of friends I have to shelter from the full reality of the situation, not so much for their sakes but for myself and Matt. Because Matt's dark moods were becoming more frequent at this time, people who cared about us began asking me more and more often, 'Is Matt okay?' I would then try to explain what we were going through and offer encouragement that he will be okay, eventually. I actually had no idea if he would be okay, but I had to believe what I told them and just keep telling myself the same thing because the alternative would have destroyed me.

Financially, we were going backwards at an alarming rate and I knew it was going to hurt us long-term if we didn't start making some drastic decisions about our future. The Army, DVA and all associated parties were still giving us hell in our search for help and Matt was silently clinging to the hope of a full recovery and that he would someday be back in uniform, which was doing more harm than good. By this stage there were more bad days than good, not only for Matt, but also for me. It felt like my world was caving in and I had been juggling balls in the air for so long that I just couldn't do it anymore. I was at breaking point and I had nowhere to turn. I couldn't see an end to it all. I was stretched so thin that I knew either something had to give, or I'd snap, and I just couldn't afford for that to happen. After all, who would keep the family together if Matt I and I were both broken? My one and only lifeline throughout this entire ordeal, at least from an ADF perspective, was Pete Goom, but he had recently been transferred to another unit and his replacement was missing in action. I don't actually recall ever meeting this person. I'm sure there was an email or a phone call once, but that was it; all the support was gone and no-one from the Army was there to help me anymore. DVA was just a bunch of people on the end of a phone sending trailer loads of paperwork for Matt to fill in and I was pretty sure they just sent document after document in the hope that they would eventually bury Matt in red tape, causing him to give up the fight as so many others had done and still do. But they underestimated his intelligence and stamina and our resolve to force people to do the right thing.

I had no other option but to remove the fog and re-set, and to do this I had to let go of something. The only thing left to give up at that point was my career, the thing that was feeding us, putting petrol in our cars, paying for insurances, electricity, the investment property in Queensland, the list goes on. But I knew leaving CHEP would mean trying to find another way to make ends meet. There was only one option. DVA had to be pushed harder to give Matt some financial aid, to give us some respite. They needed to understand that this wasn't actually their money and that Matt, as a returned serviceman, a veteran from not one but two wars, deserved better. They needed to not only help him medically, but also financially. Why should he have the stress of his family losing everything just because of their ineptitude to either get him the right treatment to heal and repair or accept that he never would? So, for the sake of my family, I made the tough call to sacrifice the only thing that was truly mine, the only thing I had achieved on my own and the one place I could escape to and just be me. Giving up my career with CHEP is one of the things I truly regret. And it made me even angrier at the Green Machine because they didn't just steal Matt's career; they took mine too.

I found out about my father's illness on the same day I submitted my resignation. It was like two worlds had collided and were crumbling simultaneously before my eyes, and there was nothing I could do but sit back and watch it all unfold.

I finished up with CHEP the same week I cremated my father. I'm sure there is irony in there somewhere.

MATT'S PERSPECTIVE

Well, the old man died. I guess it was kind of expected, but it still hit Kaz hard. As usual, she went to work organising things for everyone. I was still up in Armidale running the shop and being there for the little ones. Mum had come up to lend a hand and I felt pulled between manning the wheel at home so Kaz worried less and being with her to provide support during what was a very difficult time, to say the least. Both Kaz and I are fairly pragmatic people. We acknowledge that death is an inevitable part of life. Kaz did her thing and we put on our public faces, neutral demeanours and consoling smiles and nods to attendees at the funeral. I worried about her because she was wound up tighter than a rubber band. She hadn't really had a chance to let go of her emotions. But little by little those emotions were eventually released following the funeral and with them came whatever healing could be achieved of the tear in her heart this man left when he finally succumbed to the human condition. I guess it will be a lifelong journey, as is the case with most legacies our parents leave us with.

My in-laws and I have not always seen eye to eye, and I doubt they have ever really been thrilled with Matt's choice of a wife, although by now I'm sure they just accept it. But besides a few early hiccups and a couple of disagreements, we were getting along okay at this point. I'm not sure either sets of our parents ever really understood the depth of Matt's medical issues and what it did to him, to all of us. But I will give my in-laws this, if I needed my mother-in-law to drop everything and run, she did it without question or hesitation and Matt's dad supported it.

The accident happened in November 2010 and from March 2011 until November 2013, the DVA made Matt (and I) jump through many hoops that required us to travel all over the state and to Queensland. During this time, I'm pretty sure Matt's mum spent more time at our home in Armidale than she did at her own home on the Central Coast. She took the girls to swimming lessons, she baked with them, she played with them and she was Grandma in the truest sense to our kids and to all who entered our home. She just looked after everyone. People would rock up to our house, walk in and say, 'Hi Grandma', which she seemed to adore. She helped maintain a sense of normality for our kids through a time that was absolute hell. Matt's dad was great too because, while Grandma was with us, he was on his own at home. But he never once complained about the sacrifices they both made, well not to us anyway.

CHAPTER 22

Recognition at last

Another year was coming to an end, a year during which I had buried my father, resigned from my job and made the decision that, in order for us to survive, to really survive, I was going to have to push Matt to sell everything, make a clean break and go home. We needed to start ringing DVA every single day if that was what it was going to take. We needed to be on their case and let them know what we were enduring. We needed them to read the reports in order to understand the pressure Matt was facing every day and the turmoil he went through while the doctors played with a cocktail of medications in an attempt to stabilise him. So that is exactly what we did: we became a pain in their collective arses.

Matt is a very private person by nature, so having this constantly thrown in his face and having to constantly go over every detail and be made to feel like he had to beg DVA for financial support or aid of any kind was humiliating at best. But while it added to his anxiety and depression, he also knew that without pushing himself beyond his comfort zone, nothing would be done. So he pushed. And lo and behold, it worked! At long last we started to receive some good news from DVA, meaning we would receive some of the financial support we deserved and so desperately needed.

Towards the end of 2012 Matt finally received a partial diagnosis. I can't emphasis strongly enough how significant that was. For Matt, it meant recognition for what he was going through, not only from others, but also for himself. It gave him verification and justification. It enabled him to see how real it was; no more burying it in the sand, no more thinking it was all in his head, pardon the pun. At long last DVA were finally starting to accept liability. To reach this point was no mean feat. It was two full years, 730 days of begging, pleading, fighting, travelling to doctors and specialists in Sydney and the Gold Coast as well as to Army 'specialists' in Tamworth, Singleton and Coffs Harbour. It was a relentless, never-ending battle of being bounced from pillar to post, all while constantly begging for assistance, guidance and support. It was a cruel form of torture, but they were finally owning it, well, part of it anyway. To be more specific, at the end of 2012, DVA accepted liability for the following conditions: bruising to the left temporal scalp, a whiplash injury (which they marked as resolved), adjustment disorder – anxiety and depression, mild traumatic brain injury and post-traumatic epilepsy. This acknowledgement opened the doors for the right medical teams to work together to deliver the best treatment and support for Matt. It was a long time coming but we were finally on the right path.

MATT'S PERSPECTIVE

So, after two and a half years of doctors, reports, assessments, emails, phone calls and forms, I finally received notification that DVA had accepted liability. It felt like a win at long last, but it was actually the smallest of victories because it was simply the first step in an even longer journey. What I learnt was that liability only gives you treatment. It does nothing to shore up all the other issues, like how to pay for groceries and utilities and how to reclaim your lost earnings, and the list goes on. Well, the answer to all of those questions was, just do more of the same! We had to submit proof of the loss of earnings and a whole plethora of other items in order to try and recoup anything we had lost. And so, just as one fight ended, another was just beginning. I knew we would have to be prepared to bunker down for the long haul.

The journey wasn't over by a long shot. However, I had no idea at that point how long the road ahead was, which was probably just as well.

I was so hopeful that we would start to see a massive change in Matt. But in hindsight, I know I was also naïve. It would be another full year before we would see any real stability, and that year was filled with hit and miss attempts by medical professionals who were on Matt's side. But even they were travelling unfamiliar territory because so much of the brain remains unknown to the medical profession. Some treatments seemed to be working okay and then suddenly fail; others just sent him in a complete spin from the get-go. It was like being married to a junkie, not that I have ever been married to one, but I figure this would have been close to it. You never quite knew when a seizure would take hold, or when Matt would spiral out of control and end up in the dreaded black hole. There was no pattern or logic to any of it. It was a case of 'monitor and react', hour by hour, day by day.

While the diagnosis brought immense relief, Matt still had to come to terms with the daily realisation that he now had to live with epilepsy, a traumatic brain injury and adjustment disorder – anxiety and depression, all because a small group of people didn't do their jobs properly while he was serving our country with the Australian Army. Any one of these disorders is hard enough, but all together is devastating. Combine this with the knowledge that they destroyed the last glimmer of hope that he would ever wear the uniform again, have another go at SAS selection or even be his old self again and you would appreciate how utterly crushed he felt. But at least he was alive, and he had a diagnosis. Every win was to be celebrated.

Christmas 2012 was a little brighter than the last and the one before that. We went to Sydney and shared it with Mum. It was her first Christmas since she was about fifteen without my dad, so we rallied around her to make it special. When we came home, we made sure we had good people around us, coming and going all the time, keeping the festivities high. Matt would have times where he needed to retreat and everyone understood and let him be. We were all accustomed at this point to making allowances. We closed out the year together with good friends, temporarily putting aside the knowledge that there were some hard decisions to be made ahead.

CHAPTER 23

Goodbye ADF, goodbye Armidale

Matt and I started 2013 with the stark realisation that we needed to re-assess our long-term plans. The business had been struggling for a long time, but it was now beyond the point of repair and, as much as we had kept it going for Matt's sake, it was no longer viable to do so. We simply couldn't afford to keep the doors open anymore. I had already resigned from my job at CHEP and, ironically, after all the problems I had experienced falling pregnant with Scarlet, I discovered I was having baby number three with Matt! Remember all the heartbreak with the IVF? Remember all the guilt I carried for feeling I would never give Matt his own children? Well, here we were with two beautiful little ladies under three and a third precious daughter on the way, due in August. What a ride! We were both over the moon at the news but also concerned about having so many mouths to feed at home. I guess what rattled me the most at this time was the sudden realisation that I no longer wanted to be in Armidale. This had nothing to do with Armidale itself. We had made some amazing friends in the town who are still our friends today. The town and its people had been extremely good to us. But unfortunately, Armidale will always and forever be the place that broke my husband.

Closing the coffee shop felt like the end of an era. It was the one thing that had kept Matt going the last couple of years. I'm not at all sure how we got the formula so wrong. I guess it was just that we were totally raw in business with zero experience to draw on. And the franchise we bought into was useless at best. Add everything else that was going on in our lives to the equation and, well, it was the perfect storm for failure. We had some great times there along the way, including a number of local charity drives we got involved in. The staff we employed had become great friends of ours and our regular customers were lovely and had become part of the journey with us. Many knew about Matt's accident and the struggles he faced and would often ask how he was going and be accommodating if he was having a bad day. As for our main staff members, the phenomenal brother and sister team of Matty B and Tash, we were so lucky to have them in our corner. I'd hate to think where we would be without them.

It is no exaggeration to say that Matty B and Tash were a critical part of Matt's recovery and our story. They accepted that Matt was broken from the very start. When he was angry, they saw through it and worked around it. When he was batshit crazy, they let him be; they protected the business and the customers loved them. Matt could do as little or as much as he needed at the time and they just rolled with it. It's funny how Tash and I did not get along at first. This seems to be a theme with me; the people I hold dearest are the ones I didn't like at first. After our initial struggles, Tash and I became friends, then best friends. Matty B and Tash knew a lot about what was going on, just not the really dark bits; we kept them largely to ourselves. Sometimes Tash would sit quietly with me when she could see I was at breaking point. She just got me. She knew I wasn't a hugger and didn't like people invading my personal space, so she would just let me cry and protect me from the world for a few moments until I regained my composure.

Matty B came to work for us not long after we bought the shop. Tash joined him a while later. Between Matty B and Matt, our coffee shop became a playground of stupidity, laughter, adventure and fun. I couldn't have asked for better. Matty B and Tash had a way of helping Matt think and see things differently. On his good days they were amazing. He would come home happy, recounting the stories of the day. Even when he couldn't remember everything that had happened, he would skim over the gaps in his memory and move on because there were so many other funny stories to tell. And on the bad days, Matty

B – like Tash – just got it and took everything, all Matt's wanderings, disappearing acts and zero concept of time, in his stride.

Mathew Bender and Natasha Bender, thank you from the bottom of my heart! Despite the depth of pain and hardship we were going through, you made it a little easier for both of us. I will never complain about all the money we lost on that business – and we lost a lot – because that little coffee shop helped us save Matt. It gave us the opportunity for him to rehabilitate, to go somewhere every day where people wanted to be around him. He had friends to look out for him and we gained an extension to our family. It's how the camaraderie within the ADF should have been. Before any decision to move could be cemented, there was a major hurdle we had to get through first: Matt needed to close the door on the Army by putting his decision to leave in writing. The following is Matt's final address to the Australian Defence Force.

I am writing with regard to an incident which occurred during a training exercise with 12/16 Hunter River Lancers (HRL) at Puckapunyal Military Area (PMA) during November 2010. It is my intent to advise you of the specifics, as best I can recall, of the subsequent management of my injuries and welfare, since the occurrence of said incident.

Post-incident I was evacuated by white fleet vehicle from the night harbour location. At this time, no neck brace was fitted, nor was a complete set of stats taken, that I can recall. I was admitted to hospital at PMA Medical Centre, where a series of tests and scans were conducted, details of which are in my medical record. Upon the completion of these tests, I was advised that they revealed minimal injuries and that I had sustained a whiplash-like injury later referred to as a whiplash injury. During this time in hospital I suffered from headaches, severe pain in the neck, jaw, shoulders and upper thorax, and consistent nosebleeds which then continued for the next four or five days.

Upon receiving my scan results and the doctor stating that he could not determine any significant injury, I requested to be discharged from hospital and even contacted WO2 WB to request his assistance, as my opinion at the time was that if I was fine as far as they could tell, then I should be back in the field working. Eventually I was returned to the field and controlled pain with over-the-counter painkillers I had acquired at the on-base general store.

During the remainder of the training period, I had a pinkish discharge from my nose for several days and was in constant pain, however I was advised by medics that the pink fluid was of no concern because the doctor had released me.

Upon returning to Armidale, my employment was retained for a brief period until I could be cleared of injury. I was cleared by the Medical Officer at A Squadron and attempted to return to work. At this time, I was still in considerable pain and had begun to notice that there was a significant issue with my short-term memory. So, I requested to be reassessed and was booked in to see a civilian doctor at Peel Medical Centre. He referred me to a physiotherapist and advised me that when I returned for a follow-up consultation, he would look at referring me to a brain injury specialist.

Physiotherapy proved to help ease some of the symptoms. When I incorporated that with the assistance of a soft tissue therapist in Armidale, symptoms continued to improve dramatically, however there was still some discomfort in my upper thoracic region and I was still having cognitive issues. The civilian doctor reviewed my situation and deemed that due to the success of the physiotherapy and despite referring me to a brain injury unit, I was fit for duty and deemed to be Mec 1 (Medical Classification 1, meaning I was cleared to train for and engage in active duty).

Later, upon attending a meeting with the civilian doctor, who was not in fact a brain injury specialist, he deemed that he did not have sufficient information to refer me to any other specialist. He added that should we be able to obtain the remainder of my info, he could construct a report.

It was not until May 2011 that these documents were recovered. During that time, I had contacted both WO2 WB and Lt RL on numerous occasions in regard to my situation and was unable to find a resolution. Upon being furnished with the required documents, the civilian doctor delivered a report based on the details in the report and what he had observed during the initial consultation.

At this stage, due to my wrongful removal from the course panel, Lt RL arranged another Mec review by the Medical Officer in which I was deemed Mec 1. Therefore, I should not have been removed from the course based on this alone, especially without the consultation of a relevant SME. (All details of the later review by the civilian doctor had been removed from my file and in fact any details of this ever occurring had been deleted or were missing from files and paperwork).

Post this occurrence, I continued to request that the relevant appointments, based on the civilian doctor's assessment, be made for me in an attempt to rectify the cognitive issues I was suffering at this time. Lt RL arranged a Mec review with a third and unbiased party. At this review, neuro-psychological testing was suggested, and I was medically downgraded to Mec 301.

As a direct result of my injuries and Defence not retaining my income, my personal training business failed. My client base eroded due to my inability to facilitate them upon being cleared by the civilian doctor, which in my opinion should not have been done if there was any outstanding issue, like that of the referral to Dr MP (the brain injury specialist). I was severely financially disadvantaged due to the considerable degradation of my client base and reputation within the industry. I was also left unable to maintain insurance as I had a brain injury at the time with no consistent diagnosis. I continue to be affected by this situation as, despite entering into other business interests, my capacity for work was seriously limited compared to prior to the incident. It is my belief that I should never have been released to return to work until all issues had been addressed and rectified or an adequate plan for returning to the workforce was put in place. I questioned this several times with WO2 WB and WO1 RSM, however I was advised that I couldn't just stay on a retained income, that the budget didn't allow for it, and that they would attempt to look into things for me (this never occurred).

The management of my case has also affected my career progression. When I was withdrawn from a course I was panelled for, which was a necessity for me to maintain my current rank and vital for career progression, it had an effect on time in rank which will not commence until said course is completed.

After all this, my wife contacted the Chief of Army's office and the Minister for Veterans' Affairs. The Chief of Army responded within forty-eight hours to my wife's request for assistance.

Over the next two years, despite several meetings with 12/16 HRL and our case manager, WO1 Peter Goom, the unit still would not work within the guidelines suggested by WO1 Goom and, in many cases, did not produce minutes for meetings which laid out a plan for them to assist me and the procedures to follow to correctly assist me in returning to work. In several cases they refused to produce minutes when senior members admitted to not fulfilling their obligations and care of duty towards me. It was not until 2012 that liability was accepted by DVA and I was finally able to start getting assistance. This was directly delayed by members of 12/16 HRL who, despite numerous requests by myself, refused or neglected to provide me with relevant medical documentation.

During this entire time, my situation, my injuries and my ongoing welfare, as well as that of my family, were repeatedly dismissed or ignored. The general environment became cold and dismissive of me. And when we finally developed a return-to-work program, I was continuously pushed to fulfil more work, do more hours and work outside of my return to work strategy to the point where I needed to take time off on several occasions and I could no longer return as it was having a detrimental effect on my psychological state.

Over this time, I was emotionally, psychologically and physically affected by the actions of the representatives from this unit. It was only after I removed myself from the situation that I was able to see any dramatic improvement in my situation.

And just like that Matt's dream career was over. He was medically discharged from the Australian Defence Force and would never wear the uniform again. I wish I could say that, at this point, our nightmare was over, and we were finally able to start afresh, leaving the Defence Force behind us. But anyone who has ever been married to someone in uniform will understand when I say they never really leave; they just become ex-Army. The brain injury, the epilepsy, the depression, the anxiety and the adjustment disorder, with all the associated medication that made him go from Dr Jekyll to Mr Hyde in a heartbeat also had to remain. With doctors slowly starting to get the medication balance right and with both of us agreeing that small business was definitely not our forte, we decided to sell up everything and head home to the Central Coast. It was time to farewell Armidale.

It had been an intense fifteen years, filled with some amazing highs but also far too many lows. Armidale no longer felt like home. I had lost my dad the year before, I had walked away from my career, we had two kids under three and one teenager still at home, and now we had another baby on the way. We also had no work and a small business that had drained us financially. What's more, the journey we had been forced to take had completely worn me out. I felt trapped and I knew something had to give. While I still had the most amazing friends in Armidale – I still do – the experience was hard, the journey heartbreaking and I needed a change. I also wanted to be near Mum, who was now living alone. We had become a lot closer since Dad's passing and, even though my brother lived close to her, he was never going to be involved with her as much as we would be. It was also important to us for our kids to grow up with their grandparents on both sides nearby. Moving home to the Central Coast would tick all those boxes and allow us to start afresh.

The medication seemed to be helping to stop Matt's seizures and he was slowly growing stronger, physically and mentally, day by day. While DVA was still making him jump through hoops, they were at least starting to acknowledge him and accept some financial responsibility. But Matt didn't want to be dependent on DVA. All he really wanted was to do what he loved, what he was trained to do and what he was good at. He just couldn't imagine living the rest of his life on a medical pension. But, having finally removed the 'demons' (12th/16th) from his life, he was at least making positive progress and had begun to look beyond the sale of the business and what the move back to the Central Coast would look like. He even started to think about going back to work. What type of work was something he still had to figure out though, because a role with the Australian Defence Force was no longer an option. It took several months to organise the move. In the meantime, I was bouncing from Armidale to the Central Coast almost every week, looking at houses to buy, growing larger with each trip. Baby Adelaide arrived on the 3rd of August. She was born in Armidale and was a beautiful little bundle of joy. Now we had three little cherubs under four to look after. Thank goodness we would soon have extended family close by to help out!

MATT'S PERSPECTIVE

It was time; it was as simple as that. We had come so far since nearly losing everything. We had learned a lot, about the system, ourselves, our relationship, business, and people; but it was time to move on. We wanted to be closer to Karen's mum and my folks, especially now with three young children, so it was an easy decision to make in the end. Leaving Armidale felt like shutting the door on a dark stressful time in our lives, and we were ready. It was also going to be nice not to have to do another Armidale winter!

By November 2013, I had said goodbye to my friends in Armidale and settled the girls and me into our new home on the Central Coast. It took months to find the house and it was never actually Matt's favourite pick, but I loved it. As they say, 'Happy wife, happy life!' Matt was still based in Armidale, working on the sale of the shop, but he did come down to move everything in for me… again. We had managed to sell our house in Armidale almost immediately, so he and Ayleigh lived in a rental property in town for a few more months while everything settled with the shop and Ayleigh finished high school. Jess was already off and carving her own way through life and I had no doubt at this point that Ayleigh would soon follow suit and fly from the nest. The best you can hope for through any of it, is that you taught them right from wrong, respect, and how to work for an honest day's pay. The rest is up to them.

CHAPTER 24

Time to rebuild

As the months rolled on, things were starting to really settle down for Matt and he was feeling so good that he actually started to believe he was fine. As great as this may seem, it's also a red flag because the danger trap for anyone living with a person with his type of injuries and condition is that it's very easy to get sucked into a false sense of security with the sufferer. A little improvement can easily seem like a huge improvement. You so desperately want things to return to normal that when things are good, you relax and stop noticing the little indicators that tell you things aren't as good as they seem.

Matt and I weren't together a lot during 2013 but we would see each other as often as we could, and I would talk with him every day on video chat. But I had no idea just how well he could convince himself, let alone everyone else, how fine he was. By the time he had found a buyer for the shop and I had found us a new home, Matt had made a decision about his next chapter; he was going private. This meant he would simply be putting on a different uniform and fighting a war with a different set of rules, for private companies instead of the ADF.

There was no way he was going back to the uniform, even if he thought he was 100% recovered. He had epilepsy, so there was no way the Army would have him back – not easily anyway – and it wasn't worth the fight to get back in. The bureaucratic bullshit that went with that was way too big a hurdle for Matt to get over. In any case, they didn't deserve him. He was well and truly done with the ADF. But he couldn't see himself working as a nine to five pen pusher. So, he started researching private contracts. 'OMG, here we go again!' was all

I could think. Same work, different uniform! But the spark was back in his eye, the fire was lit and, if he could, he was going to do whatever it took to get back there, in any capacity he could, and I was excited for him. Call me naive but I just wanted him to be happy. I wanted him to have his confidence back and be self-assured and strong; I wanted it all for him, probably as much as he wanted it for himself. So, after jumping through the appropriate hoops for the private sector and going through all the interviews and assessments over a three to four-month period, he was finally offered a job.

MATT'S PERSPECTIVE

For most people, heading to an island in the Pacific, all hot, sweaty and dirty, with few amenities, and expected to work a 12/12 roster seven days a week for three weeks, doesn't seem like much chop. But for those like me, it's less about all the hardships and more about the opportunity and the ability to shine at your craft. Furthermore, the Army readies you for very specific types of relationships with those around you. There is a level of camaraderie and selflessness that can only really be understood once you have spent a significant amount of time living, breathing and, in some cases, fighting an enemy (or each other) with your workmates. There are certain things in life that are milestones and seriously impact a man's life: your wedding day, the birth of your children, when your kids go to school and joining the Army. All these moments help you grow at a macro level. At the micro level it might be simple little things like a favourite set of clothes, PlayStation with your room-mates, a tool that works perfectly, or when you get an awesome car (armoured vehicle on ops). I can honestly say that living for months at a time within arm's reach of your mates makes you part of the largest and most dysfunctional family around.

When Matt got word that he had received work in a private contract, he was elated. My God, he was like a little kid in a candy store! Finally, something was going our way. He was starting to feel whole again, like he was getting a part of himself back again. It was amazing to watch and yet sad at the same time that the Army couldn't see fit to help him get there. This change meant I was no longer an Army wife. I was the wife of a man who was broken, rebuilt and hung out to dry. He had to pick himself up, he had to work at it every single minute of every single day and, as a family unit, we had to endure everything that went with that journey. And now that he felt he was okay, he was going back, because it was all he knew; it's what gave him fulfilment.

Here's the secret to Matt and men like him. He can bury shit so deep that, on the surface, you would never know it ever existed. The sad part is, the demon never sits dormant for long and I'm not sure he will ever actually be cured. I doubt the medical profession knows enough about the brain to be able to say for certain when a problem affecting it has been solved. And a diagnosis of adjustment disorder comes with its own set of words and documents, all written about the wheres and whys but never the how to fix, just how to manage. It's such a dangerous disorder. It lulls you into thinking everything is okay; you almost get comfortable with it. And we shouldn't. We should never get so comfortable that we take the downs as normal and ignore them. Because it isn't normal.

No one actually wants these demons in their life and denial is a major factor. The beast actually thrives better if you deny its existence, ignore it or try to suppress it with alcohol or drugs. I was lucky, Matt didn't – and still doesn't – drink. But he would push boundaries like no other, as he did with his bodybuilding, for instance. It wasn't normal, but it was his way of coping, his way of feeling like he was in control, no matter how extreme, no matter how high the stakes. The faster he could drive, the more he could feel. The heavier the lift, the more pain, the more he felt. People would just write it off as determination, but it wasn't. It was Matt's drug of choice to hide from the demon and prove that he was the one in control. And all too sadly, the drug of choice becomes 'normal'. The people around them become accepting and eventually the demon strikes and wins. We should never become accepting; we should stand and fight every single day. Even when I feel like I'm breaking and I just can't see him through the bullshit and I just want to run and hide, I have to stand and fight. Never give up on them. We are all they have.

Matt eventually moved down and we were all settled back on the Central Coast, looking forward to celebrating our first Christmas in our new home and entering 2014 full of hope and excitement. This was our chance to wipe the slate clean. But it didn't stay calm for too long. Matt's new job took him to Manus Island, where he would stay on a rotating three weeks on, three weeks off roster. He was so excited; he had found real purpose again and, as far as he was concerned, he was getting back on track. It was going to be amazing. He let DVA know he was going back to work, not that they really cared. I guess for them it just meant one less lot of paper shuffling.

The Green Machine was done, and the private machine had kicked in. It's just another platform of the same shit really. I remember one time I was at home and my phone suddenly went nuts with incoming messages and phone calls. Eventually I got to my phone and checked my messages. All my friends had been watching mainstream TV ('media sensationalism' as Matt would call it) and were asking me if I had seen the news and was Matt okay? As if I had seen the news? I was too busy with three children under four! Evidently there had been an incident on Manus Island. Refugees were threatening to kill and starve themselves and the Emergency Response Team (ERT) was sent in. Guess who had been selected to lead the ERT? Yep, you guessed it: my husband! He couldn't just do normal level security, like watch a gate or sit in a tower. No, Matt Page had to be in the middle of all the action.

I figured by this time that no news from Matt or his superiors was good news. So I reported back to our well-meaning family and friends that if there was anything to be worried about, I would have been advised before the general public. At least that's one good thing about these types of institutions; they at least have the decency to remove you as the wife or a close family member from mainstream media sensationalism. They guarantee you that if anything does go wrong, they will protect you and keep you well-informed and prepared before anything hits the media. Matt eventually called and asked me if I had seen the news. I laughed and told him curiosity eventually got the better of me and I switched over to have a look. I said the only good thing to come of it was that I saw him in his riot gear. We had a bit of a laugh and he then went on to give me a little more of the story, the truth.

As a FIFO worker, time with Matt was unpredictable whenever he flew back home for his three weeks off. Most of the time we would end up in a fight and often I would be to blame because I just wanted the old Matt back. I'm not perfect. I wanted a break from Matt always fighting the demons when he was with us, I wanted to be the centre of attention for a change, I wanted our family to be normal. Sometimes I felt like I was fighting with the demon itself, not Matt. Other times I was just tired of him maintaining the charade of his 'normal' façade. So, we would fight, we would make up and we would try to ignore the demons for the time he had left with us. It never felt like three quality weeks with him at home because, aside from the fights, it would take him nearly a week just to catch up and feel almost human again. They worked them hard on Manus, twelve to sixteen-hour shifts and no real rest in between. The living conditions for the boys were pretty tough as well. At least most of them were ex-Defence, so they were used to living hard. But this was beyond that in the beginning. It took almost five months just to get them into some sort of sealed sleeping quarters.

The situation was hard on me too. While Matt's mum would help out when she could, it was largely me with three little girls all under four, 24/7, each wanting me to pay them undivided attention. I didn't like it anymore. I didn't like the separation and I didn't like the pressure of having to do it all on my own 50% of the time. And when he did come home, he would completely screw with my routine. It would be a disaster. At least when he was in the Army, he was gone for eight months at a time. Don't get me wrong; I loved our time together and always looked forward to him coming home. But the negatives were starting to outweigh the positives. I also couldn't ignore the feeling that something was wrong. Something just wasn't sitting right, and I started to see the cracks. You have to learn to be patient with men like Matt Page. You can't just charge like a bull at a gate and say enough is enough. I had to carefully bide my time. So, we kept moving forward, three weeks Manus, three weeks home.

Weeks turned into months and we were still in the same holding pattern, but it was becoming increasingly obvious that the cracks were getting bigger and they were wearing Matt down. At about the twelve-month mark, he came home looking like walking death and his head was all over the place. The nightmares, the dark moods and the depression were all starting to escalate again. I guess all sectors have their share of bureaucratic bullshit, but when you have it amongst a group of people who are risking their lives to do their job, it really becomes high risk. Unfortunately, in the private sector – this one anyway – the bottom line was more important than the boys. The job they were doing was stressful enough without egos and control issues getting in the way. It was becoming a little like the Defence Force: the wrong person in charge and all the boys were having to pick up the pieces. There were a number of boys getting out, but my husband was determined to see it through. What he didn't see was the brick wall, the demon coming straight for him at 100 miles an hour. And it hit full force. He thought he was okay. He thought he felt on top of the world and could go back to being old Matt. But you can never go back. Not ever. And the realisation of that, the realisation that it wasn't just the Army life he had lost, but that it was any form of front line fighting, the 2% line of work that very few can do – that he loved to do – was gone. His ability to operate at that level long-term was over. And over for good.

It was like going back to square one and everyday was Groundhog Day. It felt like the walls were caving in on both of us just as we were finding our way again. Nope, nope, nope! I was not going to be a victim to this hell a second time. I dug my heels in and I called it. I sat down with Matt and told him enough was enough. The madness had to stop. He needed to let go and be okay with that. He needed to go back to DVA and seek help. He had the right contacts now. He had all the case files and he knew how to get the ball rolling. I told him he had done his service and if living on a pension for a while – or even a lifetime – was the way forward, then he had better suck it up and be okay with that. I told him I would go back to work and he could stay home with the kids. I had been home nearly three years now and it was time. I didn't care what it took; he needed to be home, he needed to be working with professionals, medical and psychological. He needed to just be okay with the fact that he was broken, and it wasn't going away.

Matt lasted another six months in the job, but every trip home was worse and worse. It was a really turbulent time. He knew he was struggling and I'm sure the people he was working for would have seen the change as well, but nothing was ever said. In the end he chose us, and I am so grateful he did. When I look back at that period of our lives now, I can see that all the signs were still there, his downs were still very real. But it's as though you just get used to it and that it almost becomes a normal way of life. Matt thought it was over forever and he was holding onto that thought for dear life, telling himself, and everyone who would listen, that he was fine, that the meds were working, and that everything was great. I wanted that for him too. But I feel as though I let him down because I missed all the signs. I missed the fact that he wasn't tired from shift work and being away; he was actually depressed and in a downward spiral. And because he wasn't home three out of every six weeks, I wasn't with him enough to see the difference. When he came home, there was a distinct shift. He would tell me he was tired and that he just needed to catch up on some sleep, but he was actually hiding the truth. He was hiding it from everyone, including me, so I didn't see it. I guess he didn't want to go backwards and that's why he panicked and hid the truth.

I often tell people my husband is broken; he is a broken soldier. For most, that might seem like a really negative thing to say about him. People even tell me I should think it, but not say it. Well, here's a news flash. I'm not ashamed of him being broken. He didn't do this to hurt his family. He didn't ask the driver of the Bushmaster to drive it into the ditch. He didn't ask for the reservist unit to let him down. He didn't ask the Army doctors not to take him seriously. He didn't ask for any of it. This is him now, this is who he is. If he had broken his spine and became paralysed from the waist down, we would say he was a paraplegic. If he lost a leg, we would say he was an amputee, but a brain injury isn't so neatly boxed. He is a broken soldier; he is my broken soldier and I'm proud of him and the journey he has endured. He still smiles and lights up my world. I'm not ashamed to say he is broken, nor am I putting him down. People shouldn't be ashamed of depression, anxiety or mental illness of any kind. People should be able to say they are broken, and be open about needing help, just as cancer patients need chemo or people with a broken bone need plaster. Matt should absolutely be proud of the man he is today, just as we all are of him. He has had to endure more than a soldier on home soil ever should. Even when he is suffering, he still gets up every day and tries to make people smile and laugh. Matthew Page is one of the most generous people I know. I am so very proud of my broken soldier.

If you are reading this and suffer from PTSD, adjustment disorder or any other debilitating mental illness (or live with someone who does) I really hope you know you can't control it – not with alcohol, not with illegal drugs, not by ignoring it and certainly not by yourself. You may be able to manage it – to a degree – but you can't control it. It's okay to let people in and to reach out for help. I have watched the demon take control of others and they tried to dowse it with alcohol. Very few are still here to tell their story (thank God, Matt doesn't drink). Here is the cold hard truth: death is guaranteed to us all. It's the experiences you have and share with your loved ones during the time you have that count. Please get help if you are struggling or live with someone who is. Please don't let the demon win by drinking or taking uppers and downers. Seek the right help, and if you are unsure what the right help looks like, then ask; ask anyone who will listen.

CHAPTER 25
Same same but different

It took a lot for Matt to finally acknowledge that his injuries were not going to heal in the foreseeable future and that the demon would always lurk in the corner, although not always within his direct sight. This was our new normal and we had to accept that this was how we needed to live for the foreseeable future. In early 2015, Matt got the ball rolling to have DVA take over his case again. After plenty of paper shuffling, more doctors' appointments and countless phone calls, he finally got the answer he needed: DVA was willing to review his case again. He got the pension sorted out and took permanent leave from his job on Manus Island. He was now officially a medically pensioned ex-soldier.

He became a house dad and, to be honest, our three little ones couldn't have been luckier. From that moment on, their dad became their Peter Pan and they his Lost Boys (girls). It was time for me to return to work and I found a job close to home doing a very similar role to the one I did with CHEP, just for a much smaller company. Our lives had finally started to settle. We weren't rolling in money, but we were okay. We had a nice home and we were reconnecting with old friends and making some great new ones, building a life that was as close to normal as we could get. The kids had settled into school and day care and the two older girls were finding their own feet in life. Ayleigh had even decided to join the Army! And with our full support too. Before you gasp in shock and wonder if we'd both lost it, Matt and I firmly believed it was her journey and her decision to make. What's more, there was never a day that we didn't believe in the Defence Force. I will always stand up for the soldier prepared to put on a uniform and do what most can't and would never choose to do. Their service is what gives me freedoms, like writing this book. I believe we need a strong Defence Force, but I also believe that individuals can effect change, given the right platform. And I strongly believe our Defence Force and Department of Veterans' Affairs are ripe for change. They need new blood; they need to be better at preparing our soldiers and their families for war and they need to be better at supporting the ones not strong enough to make it to war and also those who do and then break. We all need to be better at supporting them. So yes, I fully supported Ayleigh's decision to join the Australian Defence Force. Unfortunately for Ayleigh however, her days in the Army were short-lived. After discovering a tumour in her leg, she was medically discharged, ending her Army career, at least for now. But that it is her story to tell, not mine.

Matt was back with all the doctors trying to get his condition under control again. For some reason, perhaps due to all the upheaval of the last six months, his epilepsy medication had ceased to work, and the seizures had returned. Silent seizures suck. He had become very good at hiding them, so it took a little while for me to realise he was having them again. He was so desperate to live a normal life that he really struggled whenever one of his conditions flared up again, as he saw it as a threat to the new life he was building. He also didn't want me worrying, but that was a Catch-22, because whenever his actions began to change, I knew something was wrong, so naturally, I worried. There was always this need in Matt to hide it from the world, me included. Was it shame? Was it denial? Or is it just the way we are as a society when it comes to mental health, and even conditions like epilepsy? I found myself asking, 'Do I support him the right way?' 'Do I say the right things?' I just wish there was a rule book for partners of current and former servicemen and women. I often wonder how different the journey would have been had the culture in the Army towards invisible ailments been more supportive.

No soldier should ever be in a position where they feel it would have been a better outcome had they been brought back from war in a box. Yet that was Matt's perception all too often. And sometimes – even today – in his dark moments, he still believes we would all have been better off if he had been killed on the front line while serving in Afghanistan or Iraq. While I know that is the demon of depression talking, it is mixed with the Army's failure to provide adequate support. Matt had to learn to deal with these thoughts, combined with the fact that constant nightmares screwed any chance of a decent night's sleep. More often than not, he was up walking the floors at all hours of the night, unable to still the mind. No sleep meant he was more irritable during the day and that has always had a negative effect on his state of mind and other ailments. It became a vicious cycle.

After much trial and error, the doctors eventually found the right formula to return Matt to some sort of stability and, by the end of 2015, he was finally beginning to come to terms with his condition. I think, for the first time since the accident, he understood that this was how life would be from now on; this was forever.

I think we both came to the realisation at that point that it was very unlikely he would ever be able to hold a stable job again. A nine to five job would mentally do him more harm than good and the thing he loved, the thing he excelled at, was lost forever.

He was now a full-time stay-at-home dad, with three very bright cherubs, who adored and idolised him. He had a support network of doctors and counsellors around him and was working regularly with them.

Unfortunately, nothing had changed with DVA. Every time Matt needed to adjust something or seek further assistance, every individual representing DVA would make Matt feel like it was money coming from their own back pocket. Nobody deserves to be treated like that, like they have to beg for assistance. I hear time and time again, through our circles of broken soldiers and their families, how hard DVA makes the process to get the right help. Why? Where does that mentality come from? Have these men and women not endured enough? I agree there needs to be a formal process, but let's go into that process with the aim of making it feel like you are actually trying to help them DVA, not make them feel like bludgers or criminals, making them have to fight to prove over and over and over again that they are broken and genuinely need the assistance. As a taxpayer, I know where the money for these broken soldiers comes from: our tax. Why is it that politicians can earn six figure incomes and receive travel privileges and God only knows what other perks, including massive super contributions from the taxpayer, but our mentally broken soldiers have to fight tooth and nail for help. Where is the Aussie spirit of fairness in that equation? When will this ridiculous attitude of 'If we can't see it, it mustn't be real' be done with? Mental health is not tangible. You will never hold it in your hand. You cannot band aid or plaster it, operate on it or stitch it up. But that doesn't make it any less real. How much more did 'the powers that be' expect them to bear? Their soul is already broken. What's left?

Matt and I continued learning and managing our new roles as best we could. Willow and Scarlet were now settled in school and baby Adelaide was in day care. As much as Matt had settled into his new role, and, to a certain degree, had accepted his conditions, he was also becoming increasingly bored, especially when things were stable, lulling him into a false sense of security. So, he decided he wanted to study and get a degree, and not just any degree, but an MBA. Furthermore, he wanted to complete it in just twelve months while continuing to look after things at home! Now, by this stage I had learned to accept that when Matt sets his mind to do something, he is so intensely focused that he is like a dog with a bone. He applied, was accepted and, true to his word – and to his credit – completed all eleven subjects in eleven months. By the time he had finished his dissertation, it had taken a little longer than his goal of twelve months. So instead of graduating in April 2017, he graduated with his Master of Business Administration in November 2017. I could not have been prouder of him.

CHAPTER 26

The next chapter: HOPE

It is now 2020 and, as I watch the world unfold with new disasters that will forever change the way we live, I am finally drawing a close to this book that has taken me three solid years to write. Along the way I have cried, I have screamed and sometimes I got to have a laugh. But most importantly, I have healed. Matt and I are closer than ever, and our small family unit is doing just fine. We are still on the Central Coast; I'm still working for that little company – although it's not so little now – and the girls are all growing way too quickly. Willow is nearly eleven and it has been ten years since the accident.

As I look back now, I'm amazed at my husband's strength and stamina. Every year since the fight from Hell that started on November 19, 2010 has brought trials and tribulations. For seven full years we fought, sometimes taking a step forward, other times ten steps back, to help my husband heal, to help him come to terms with his new way of life, for all of us to come to terms with our new way of life. I know our story is just one of so many and I also know firsthand that there are many others still fighting the battle, some who have been fighting it a lifetime longer than we did. We are truly a success story. For all the tears, for all the hurt and all the pain and suffering, we survived the inept attitude of the Australian Defence Force. And truth be told, if Matthew John Page had his time over, he would join again, despite the system and the people who failed him.

I don't wish our story on anyone. We need to find a way to do better for the broken. It's a sad state of affairs when, in today's society, we are more worried about being politically correct, than calling a spade a spade. How many soldiers have to kill themselves because they feel there is no hope? We will most likely always live with the black demon from Hell sitting forever in the corner of Matt's world, and there are times I can catch him before he falls and other times I miss the signs and we have to fight together to dig him out and get him back on track. Matt still gets bounced from medication to medication and shuffled from doctor to doctor, even now, ten years on. The only thing that has changed is how Matt and I approach it, but the process itself has not improved.

There are times today when I think the bad times are well and truly behind us and then, BANG, just like that, I am reminded that we must always remain on high alert. I can never take my finger off the pulse, I can never just relax and think he is cured, because no matter the medication, no matter the people charged today with helping him/us, there will always be the next black hole that he will fall into, and I have to be ready. That has been a hard lesson to learn. Things will never go back to the way they were before the accident, but in so many ways, life is better than it was because we were forced to learn so much about ourselves and each other and discover new things to be happy about and thankful for. We grew so much as human beings.

I am far better these days at helping my husband when he falls, and to his credit, Matt is so much stronger, so much more open to his conditions and willing to fight and beat the demon, now that he understands it more. I don't know what it all means for us long term. I don't know if Matt will ever get so low again that we will lose him. I believe this will forever be a fear tucked in the corner of my world. Mental illness is such an unknown entity in any one's world. The best I can do is live one day at a time, enjoy the good times and dig deep during the bad times, hoping that at some point in our lifetime, the so-called experts will find the key that unlocks the secrets to this elusive illness.

No one can change what we went through, but I hope that, at the very least, we can learn from it. As a human race we will always have conflict, there will always be threats of war just over the horizon or on our own soil and, while we may say we don't want war, sadly conflict is inevitable and the need for soldiers will always remain. But we also need to acknowledge all the other work the defence forces do, like international humanitarian work, national support work, the list goes on. And to be able to continue this work, to be able to be ready to protect our freedom if and when the need arises, we must have the right people in place, trained and ready for action. We need a strong and viable Defence Force with the right processes and support structures in place to help when things go wrong. I believe we will get there one day. For now, if you are the sufferer, there are places and people you can turn to; there is light at the end of the tunnel, no matter how far away that may seem. Please know you are worth saving and people are *not* better off without you, ever. If you are the support person to the sufferer, know that you are also worth it, that the fight is hard but the person you are fighting for is worth the journey. Don't be too afraid to reach out for help and please open your doors to let others know how much you are hurting.

On ANZAC Day and Remembrance Day and any day where we, as a nation, come together to commemorate and say thanks to our armed servicemen and women, past and present, please remember that behind every single person in uniform – any uniform – is a family willing their loved one to come home. For every broken soldier there is a wife, a husband, partner or family member, doing their best to hold it all together and try and help, hurting right alongside the sufferer. It's about finding the right support structures that help them all. We can all do so much better at understanding and acknowledging that.

This is just my story, but I know there are so many more, hundreds if not thousands more untold stories that deserve to be heard. My healing was writing this book, my way forward was getting it all down on paper. I started this book in anger, wanting answers and revenge for the pain we endured, but I've completed it thankful; thankful we were strong enough to endure, thankful that we are still here, still together, watching our kids grow up, building our lives in a place where our days are filled with more joy and happiness than darkness, demons and desperation. I'm thankful that we are in a position where we can help others and make a difference, no matter how small. Our family motto has always been: Smart people work hard every day'. But we have added to that now: HOPE – Help One Person Every Day.

Made in United States
North Haven, CT
04 November 2026

21760246R00127